Laboratory Studies of Chick, Pig, and Frog Embryos

4th Edition

Ray L. Watterson, Ph.D.
University of Illinois
Urbana, Illinois

Gary C. Schoenwolf, Ph.D.
University of Utah
College of Medicine
Salt Lake City, Utah

Robert M. Sweeney, Ph.D.
(Deceased)

Burgess Publishing Company
Minneapolis, Minnesota

COVER: 33-hour chick embryo whole mount.

Printed in the United States of America
ISBN 0-8087-2373-1

0 9 8 7 6 5 4

Contents

Preface

In the fourth edition, the text has been thoroughly revised and condensed by Drs. Schoenwolf and Watterson. Most of the photomicrographs have been relabeled to increase clarity and to correspond to the revised text, almost all text figures have been modified, and several new figures have been added. These changes will considerably reduce the time required for detailed laboratory study of vertebrate embryos. We believe that these changes will also increase the student's comprehension of the complex morphogenetic events occurring during early vertebrate development and allow students to study these events more independently.

As in previous editions, students are first directed to the detailed study of the 33-hour chick embryo. We are convinced that this is the best way to introduce students to the study of vertebrate embryos. By studying the 33-hour chick embryo, students learn (1) how to use whole mounts and serial sections to study embryonic stages, (2) how to visualize structural relationships between parts of a relatively simple developmental stage, and (3) the basic terminology required to describe earlier and later developmental stages. Following study of the 33-hour embryo, students examine *earlier* stages in detail to learn how the structural relationships in the 33-hour chick embryo originated. They then study *later* stages in detail to learn how the structural relationships in the 33-hour chick embryo are progressively modified during further development.

A detailed but clarified description of the 10-mm pig embryo is provided for the study of mammalian organogenesis. In addition, a new section on early stages of frog development has been added, and the entire section on frog development has been revised so that students may begin their study of vertebrate development with either the 33-hour chick embryo or with the early frog embryo. To aid students in understanding the relationships of serial transverse sections to whole mounts, drawings of whole mounts indicating the levels of specific sections have been added to plate legends. The index at the back of the manual lists all terms designated in the text by ALL CAPITAL LETTERS.

In preparation of this edition, appreciation is due to Sister Renée Costanzo and Mr. Maurice Pescitelli for their suggestions. We especially acknowledge our indebtedness to Mrs. Alice Prickett, for cooperation and expertise in drawing the revised and new text figures; to Mrs. Patricia Schoenwolf, for assistance with proofreading and preparation of the index; and to Ms. Anita Kimbrell for secretarial assistance.

Finally, it is our sad duty to inform our readers of the untimely death of Dr. Robert M. Sweeney. His many contributions to previous editions of this manual have been largely responsible for its success. To his memory the present edition is proudly dedicated.

Gary C. Schoenwolf
Ray L. Watterson
September, 1978

I

Introduction

Laboratory work in introductory courses in development usually includes studies of *developmental anatomy.* Suppose you obtained some preserved frog embryos at the stage illustrated by Fig. Aa and wanted to study their anatomy. Under the microscope you could identify a few poorly defined external features, as well as the body axes (*cranial–caudal, dorsal–ventral, right–left).* To study internal features in detail it would be necessary to slice (section) the entire embryo into thin sections of a given thickness. One or more slides containing every section from the first one (most *anterior* one) to the last one (most *posterior* one) is called a *set of serial sections.* The most anterior section of this set (#1) is mounted at the upper left-hand corner of the slide, and successive sections are mounted in the following way:

1	2	3	4	5	6	7	8	9	10
11	12	13	14	15	16	17	18	19	20
21	22	23	24	25	26	27	28	29	30

If there are too many sections to mount on one slide, the more posterior ones are mounted on slide #2 of the series in the following way:

31	32	33	34	35	36	37	38	39	40
41	42	43	44	45	46	47	48	49	50
and so forth.									

Fig. Ab illustrates four representative *transverse sections* (that is, sections cut perpendicular to the cranio-caudal axis). Fig. Ac illustrates these same sections after they have been transferred to a glass slide and mounted from left to right in the order in which they were cut. Exactly how much of the anatomy of the embryo can one expect to see in any one representative section? Suppose that your set of serial sections contained a total of 100 sections. The sixth section of the set might cut through the level of the developing *eyes;* in such a section you could determine the relationship of the eyes to other structures (Figs. Ab, Ac). After identifying the various structures at this level of the embryo, you might then examine more posterior sections, such as #15 through the *heart,* or still more posterior sections (#50, #80). Unfortunately the study of individual sections provides only a two-dimensional picture of the embryo. To understand the anatomy of the embryo in *three* dimensions you must visualize each section as again part of the whole embryo. For example, the *notochord* can be identified in sections #15, #50, and #80. By connecting the section of the notochord at the level of section #15 with the section of the notochord at the next level (section #50) and those at successive levels, you get an accurate picture of the cranio-caudal extent of the notochord as well as its relationship to other structures. In the same way you can get an accurate picture of the cranio-caudal extent of the *neural tube (brain* and *spinal cord)* and *digestive tube,* as well as their relationships to each other and to other structures. *The most difficult task facing the beginning student is to learn to visualize relationships of parts of an embryo to one another in three dimensions.*

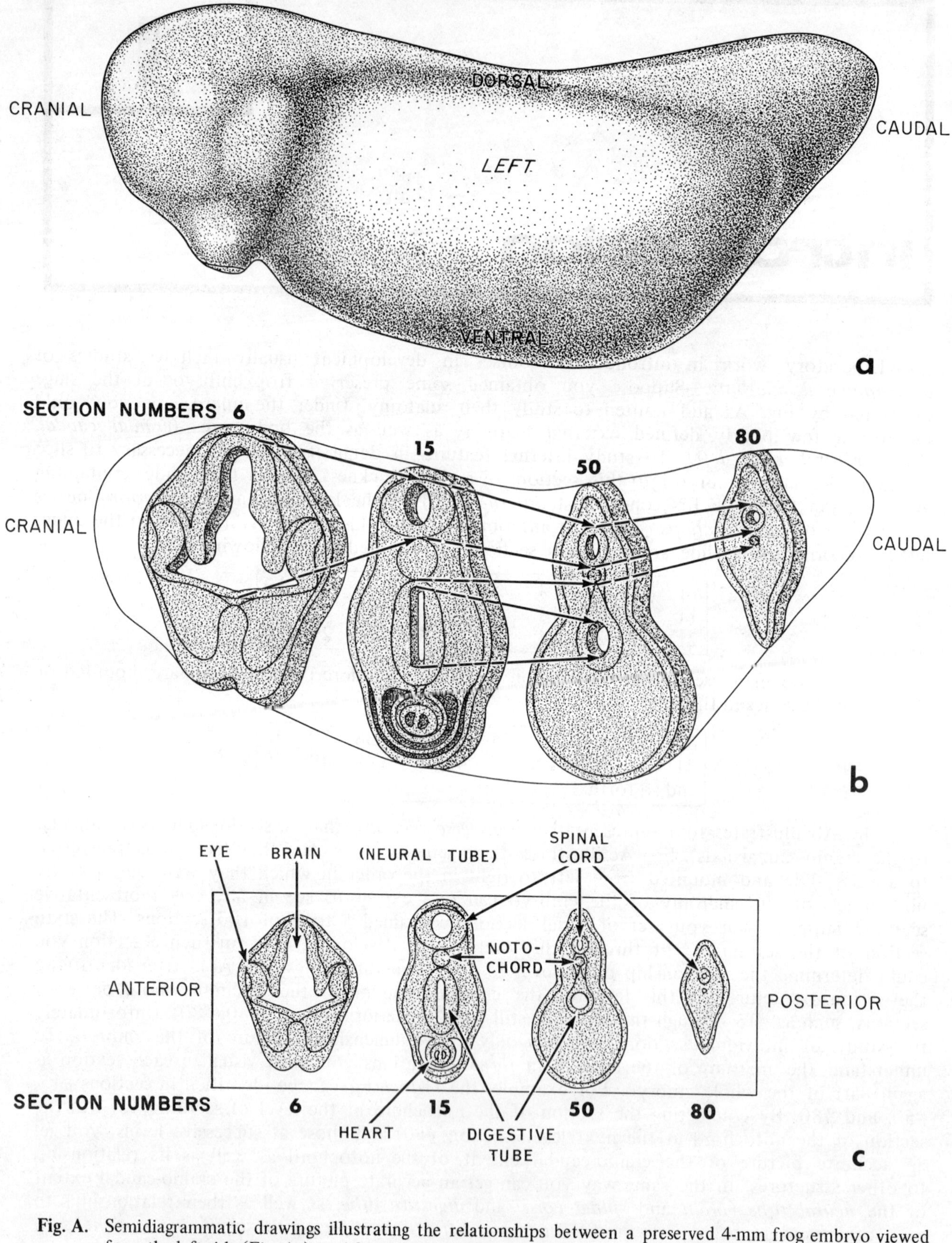

Fig. A. Semidiagrammatic drawings illustrating the relationships between a preserved 4-mm frog embryo viewed from the left side (Fig. Aa), and four representative transverse sections through this embryo (Figs. Ab, Ac).

II

Useful References

Although this manual is thoroughly illustrated by photomicrographs and line drawings, certain textbooks provide additional helpful illustrations. Each textbook will be referred to by the letters in parentheses listed below. Because textbooks vary in terminology, the student should follow the terminology used in this manual to avoid confusion.

(A) Arey, L. B. 1974. *Developmental Anatomy*. Rev. 7th ed. Philadelphia: Saunders.

(B) Balinsky, B. I. 1975. *An Introduction to Embryology*. 4th ed. Philadelphia: Saunders.

(H) Hamilton, H. L. 1952. *Lillie's Development of the Chick*. New York: Holt, Rinehart and Winston.

(HU) Huettner, A. F. 1949. *Comparative Embryology of the Vertebrates*. Rev. ed. New York: Macmillan.

(M) Mathews, W. W. 1976. *Atlas of Descriptive Embryology*. 2nd ed. New York: Macmillan.

(P) Patten, B. M. 1971. *Early Embryology of the Chick*. 5th ed. New York: McGraw-Hill.

(PA) Patten, B. M. 1948. *Embryology of the Pig*. 3rd ed. New York: McGraw-Hill.

III

Early Embryology of the Chick

The chick embryo has long been used for the study of early stages of embryonic development because the developmental stages desired can be readily obtained. Many of the developmental changes in the chick embryo are almost identical with those characteristic of other vertebrates, and particularly of mammals, whose young embryos are difficult to obtain in sufficient numbers in the stages needed for study. Consequently, the chick will be used extensively for the study of early stages of development. The stages selected for study are those that have been designated classically as 18-, 24-, 33-, 48-, and 72-hour embryos. Embryos within each classical stage vary somewhat in the degree of development attained and thus your embryos may be slightly different than those described and illustrated here. Descriptions have been sufficiently generalized to permit study of most embryos within each classical stage.

A. THE PART OF THE EGG TO BE STUDIED

From an embryological point of view the bird's egg consists of a large spherical mass of YOLK on which rests a small white disc of cytoplasm. If this disc has not initiated CLEAVAGE (that is, is unsegmented), it is called the BLASTODISC. If this disc has initiated cleavage (that is, is segmented into two or more cells by the process of mitosis), it is called the BLASTODERM. All parts of the embryo and its membranes originate from the blastodisc.

Developmental stages can be examined in detail only after the blastoderm is removed from the yolk and then either mounted flat on a slide so it can be studied as a whole mount or cut into serial sections and then mounted. The center of the blastoderm is separated from the underlying yolk by a SUBGERMINAL CAVITY; the periphery is not. Therefore, when the blastoderm is removed from the yolk surface, its center is more translucent than its periphery to which yolk adheres. The central region of the blastoderm is therefore named the AREA PELLUCIDA, the peripheral region the AREA OPACA.

B. 33-HOUR CHICK EMBRYOS

1. **Whole mount. Aids:** Fig. B; Plate 1; A, Figs. 543, 544; B, Fig. 252; H, Figs. 67-70; HU, Figs. 105-107; M, Figs. 92, 94, 95; P, Figs. 53, 55, 57, 64.

First examine your slide with the naked eye. Note the round to oval shape of the mounted specimen. The specimen is divisible into two areas, the peripheral AREA OPACA and the central AREA PELLUCIDA (Fig. B). The area opaca is further divided into the inner mottled AREA VASCULOSA, containing developing BLOOD ISLANDS, and the outer AREA VITELLINA, a nonmottled area overgrowing the yolk. Try to identify the SINUS TERMINALIS, a circular blood vessel separating the area vasculosa from the area vitellina. This vessel may appear light or dark depending on whether it contains blood cells. Most of the area vitellina is usually trimmed

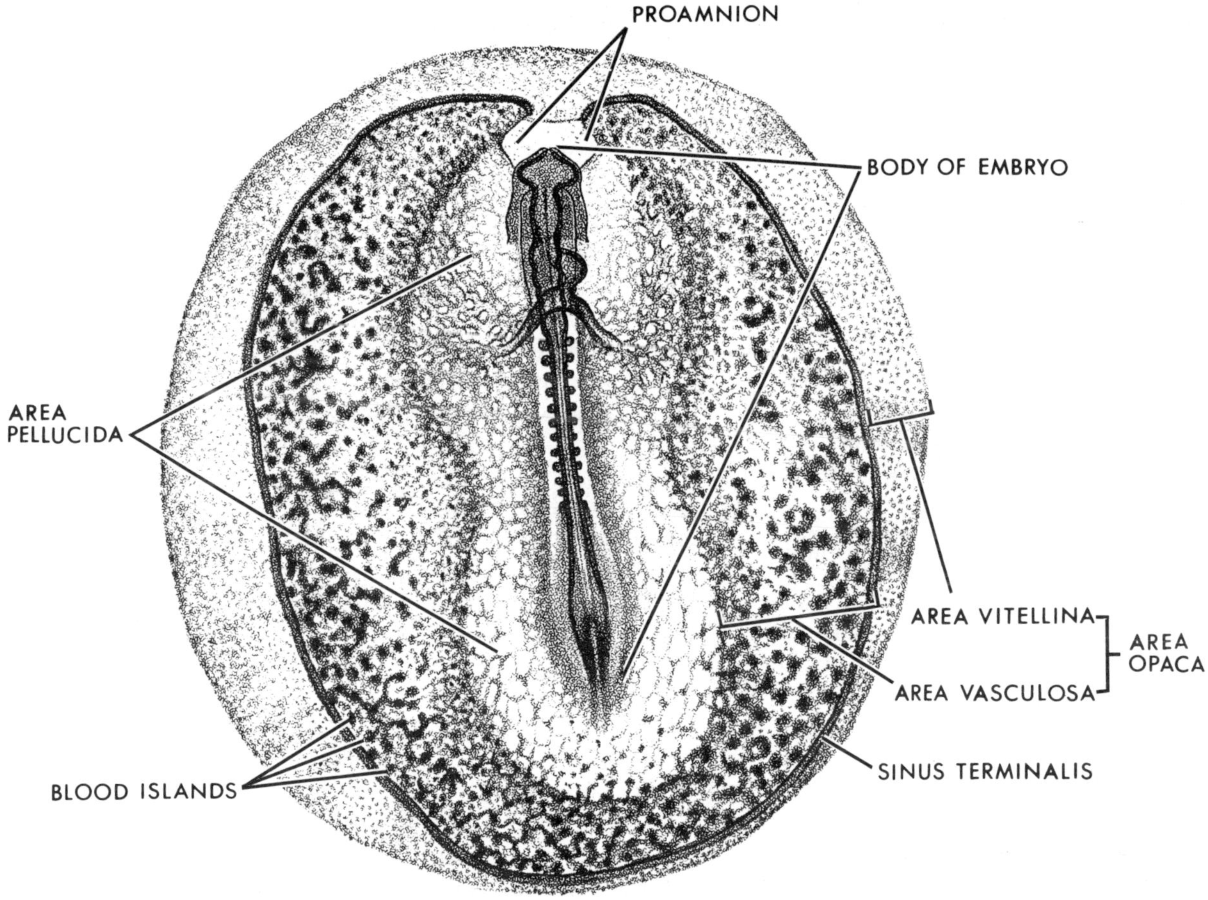

Fig. B. Drawing of a 33-hour chick embryo whole mount.

off from mounted specimens, so the sinus terminalis will probably be located just within the margin of the blastoderm. Running lengthwise within the area pellucida is a dark region, the BODY of the developing embryo. The darker and wider part of this region is the HEAD (CRANIAL) END of the embryo; the lighter part is the TAIL (CAUDAL) END. Note a very clear area beneath and in front of the head; this area is called the PROAMNION. It is the only region of the blastoderm which still lacks the middle GERM LAYER, the MESODERM.

Position your slide on the microscope stage so that when viewed through the microscope the embryo is oriented as in Plate 1. Examine the slide under low magnification. First examine the AREA VASCULOSA. The BLOOD ISLANDS, which give this area its mottled appearance, consist of irregularly shaped accumulations of mesodermal cells. The peripheral cells of each blood island form the ENDOTHELIUM constituting blood vessel walls; central cells form PRIMITIVE BLOOD CELLS. Note that blood islands are most numerous lateral and caudal to the body of the embryo, and that they are fusing together to form blood vessels. The primitive blood cells are therefore enclosed within these developing blood vessels, the VITELLINE BLOOD VESSELS, which may open peripherally into the sinus terminalis, and which may extend across the area pellucida towards the body of the embryo. Note that within the area pellucida blood vessels are clear (empty) channels containing no blood cells, and that the vitelline vessels are all about the same size. Although the heart at this stage contracts weakly, circulation of the blood begins slightly later. Initiation of the heartbeat and onset of circulation begin long before the heart is supplied with nerve fibers.

Next examine the HEAD END of the embryo. Focus up and down on the head with the coarse adjustment of your microscope. Different structures and lines will become apparent to you at different levels of focus. This means that in addition to *length* and *width,* the body of the embryo has considerable *depth;* therefore you can focus only upon a part of its thickness at one time.

Notice two dense, thick bands running quite parallel to each other throughout the middle region of the embryo (Plate 1, Figs. 1-3). Just lateral to these bands are paired blocks of mesoderm, the SOMITES, separated by INTERSOMITIC FURROWS. These bands are the thick lateral walls of the NEURAL TUBE. The neural tube is formed from the outermost GERM LAYER, the ECTODERM. Cranially, where the bands diverge in a bilaterally symmetrical fashion, they constitute the lateral walls of the BRAIN. Caudal to the brain these bands constitute the lateral walls of the SPINAL CORD. Thus most of the CENTRAL NERVOUS SYSTEM (that is, brain and spinal cord) present at this stage exists as a *hollow ectodermal tube.* More caudally the bands diverge laterally as NEURAL FOLDS. The space between the neural folds is the NEURAL GROOVE. At the caudal end of the embryo the converging bands represent the PRIMITIVE RIDGES, which are the lateral margins of the PRIMITIVE STREAK. At this stage the primitive streak is the site of mesoderm formation.

Turn your attention now to the brain region and note the two largest lateral expansions (that is, evaginations) of the neural tube; these are the OPTIC VESICLES, the beginnings of the EYES. The level of the brain from which these evaginate is the DIENCEPHALON. By careful focusing, note that these vesicles are in close contact laterally with the overlying SKIN ECTODERM. These regions of skin ectoderm are probably not thickened, but they will *later* thicken and transform into the LENSES of the eyes. Experiments have shown that the optic vesicles stimulate (that is, induce) the ectoderm to form the lenses. Note a dark, shadowy mass lying in the midline between the optic vesicles. This is the INFUNDIBULUM, which forms as an evagination from the floor of the diencephalon (Plate 1, Figs. 4, 5). It will *later* form the POSTERIOR PITUITARY GLAND (NEUROHYPOPHYSIS).

The level of the brain cranial to the diencephalon is the TELENCEPHALON. At this stage it is poorly demarcated from the diencephalon and the two are usually spoken of collectively as PROSENCEPHALON. Focus down into the telencephalon and notice a white median notch or cleft, the CRANIAL (ANTERIOR) NEUROPORE, through which the cavity of the neural tube is still open to the outside.

The division of the brain caudal to the diencephalon is the MESENCEPHALON, and the next caudal expansion is the METENCEPHALON. The most caudal region of the brain is the MYELENCEPHALON, which is characterized by several smaller enlargements of the neural tube, the NEUROMERES. The metencephalon and myelencephalon are spoken of collectively as the RHOMBENCEPHALON. Note that the space between the walls of the brain and the skin ectoderm is filled with loosely packed cells, the HEAD MESENCHYME.

Recall that the somites are located lateral to the neural tube throughout the middle region of the embryo. How many somite pairs are present in your embryo? The most cranial pair of somites is incomplete (has no definite cranial boundaries). It lies just caudal to the last neuromere. All somites cranial to the fifth intersomitic furrow are later contained within the head of the chick. Thus more than one-third of the embryonic body formed at this stage is future head. Note that lateral to each neural fold a band of tissue extends caudally from the last pair of somites. These two bands are the SEGMENTAL PLATES, which will form additional somites in cranio-caudal sequence. By carefully focusing down through the neural tube you will be able to identify a narrow mesodermal rod, the NOTOCHORD. This structure lies directly beneath the neural tube and neural groove, between the infundibulum and the cranial end of the primitive streak. The cranial tip of the notochord induces the infundibulum to form. Notice that the notochord is widest just cranial to the primitive streak.

By carefully focusing just lateral to the mesencephalon and rhombencephalon, lines extending cranio-caudally can be observed. These lines represent the lateral boundaries of the FOREGUT (see H, Fig. 69; HU, Figs. 105, 106; M, Fig. 94; P, Fig. 55). The cranial boundary of the foregut usually lies near the caudal boundaries of the optic vesicles but it is often very difficult to see. The foregut is formed from the innermost GERM LAYER, the ENDODERM, and resembles a finger of a glove, closed cranially but open caudally. The opening is called the CRANIAL (ANTERIOR)

INTESTINAL PORTAL (Plate 1, Fig. 4) and can be seen more readily if slides are viewed from the ventral surface (Plate 1, Fig. 5). It lies beneath the myelencephalon and its margin forms a distinct curvature with its concavity facing caudally.

Note that the cranial part of the head is sharply bounded and that it protrudes above the underlying blastoderm. The head is separated from the blastoderm by an ectoderm-lined space, the SUBCEPHALIC POCKET.

Focus down through the level of the rhombencephalon and observe the HEART, usually bent somewhat towards the right. It can be seen more readily from the ventral surface. Entering the heart from behind, immediately cranial to the cranial intestinal portal, are two large veins, the VITELLINE VEINS. The cranial ends of these veins are continuous with the SINO-ATRIAL REGION of the heart (*future* SINUS VENOSUS and ATRIUM). That region of the heart which bulges slightly to the right (or the widest part of the heart if bending has not yet occurred) is the VENTRICLE. Cranial to the ventricle the heart narrows as the BULBUS CORDIS.

2. Serial transverse sections.

*a. **Instructions for preparation of graphical reconstruction.*** To help you visualize relationships of parts of an embryo to one another in three dimensions, it is recommended that you prepare a graphical reconstruction of *your* embryo. Examine your slides with the naked eye. Note that the sections are mounted in horizontal rows. Count and record the number of rows and the number of sections in each row. Add the number of sections in each row to determine the total number of sections.

On a sheet of graph paper ruled 10mm/cm (Fig. C, only centimeter lines indicated) let each millimeter space along the longest (horizontal) axis represent the thickness of one section. At the top of the paper number off every 10 sections from left to right. Then draw vertical dashed lines at the levels of the last section number in each row. Number the rows at the bottom of the paper. At the left subdivide your paper into three areas by brackets enclosing the *exact* centimeter lines shown in Fig. C and label these areas as indicated. Within each subdivision, place the following labels at the right of the page. Under *ectodermal structures* label the first centimeter line from the bottom *neural tube and groove,* and the third centimeter line *subcephalic pocket.* Under *mesodermal structures* label the first centimeter line *notochord,* and the second centimeter line *head mesenchyme, somites, and segmental plates.* Label the space between the fifth and sixth centimeter lines *heart and related vessels.* Label the ninth centimeter line *right dorsal aorta.* Under *endodermal structures* label the second centimeter line *foregut.*

*b. **Study of slides. Aids:*** Plates 2, 3; B, Figs. 255-261; HU, Fig. 108; M, Figs. 96-103; P, Fig. 63. Position your slide on the microscope stage so that when viewed through the microscope each section is oriented as in Plate 2 (that is, with dorsal surface at the top). *Do not place microscope stage clips on slides to hold them in place as this could fragment sections!*

Begin by examining sections in antero-posterior sequence (see *Introduction*) until you find a section resembling Plate 2, Fig. 1. This section cuts through the portion of the head containing the DIENCEPHALON. The PROAMNION lies beneath the head, separated from it by a space, the SUBCEPHALIC POCKET. The proamnion consists of two layers: upper ectoderm, lower endoderm. To either side of it are four cell layers. The dorsal two constitute the SOMATOPLEURE (upper ECTODERM, lower SOMATIC MESODERM), the Ventral two the SPLANCHNOPLEURE (upper SPLANCHNIC MESODERM, lower ENDODERM). The space between somatic and splanchnic mesoderm is the COELOM.

From this level trace sections *anteriorly* until the head disappears. Identify the proamnion, somatopleure, splanchnopleure, and coelom at this level. Then trace sections *posteriorly* and note the reappearance of the head. The first part of the brain seen is the TELENCEPHALON. [Record the first section through the neural tube on the *neural tube and groove* line of your graph.] It will usually appear divided into right and left halves by a narrow white slit, the CRANIAL NEUROPORE.

Continue to trace sections posteriorly. The region of the brain to which the OPTIC VESICLES attach is the DIENCEPHALON (Plate 2, Fig. 1). Part of each optic vesicle may lie

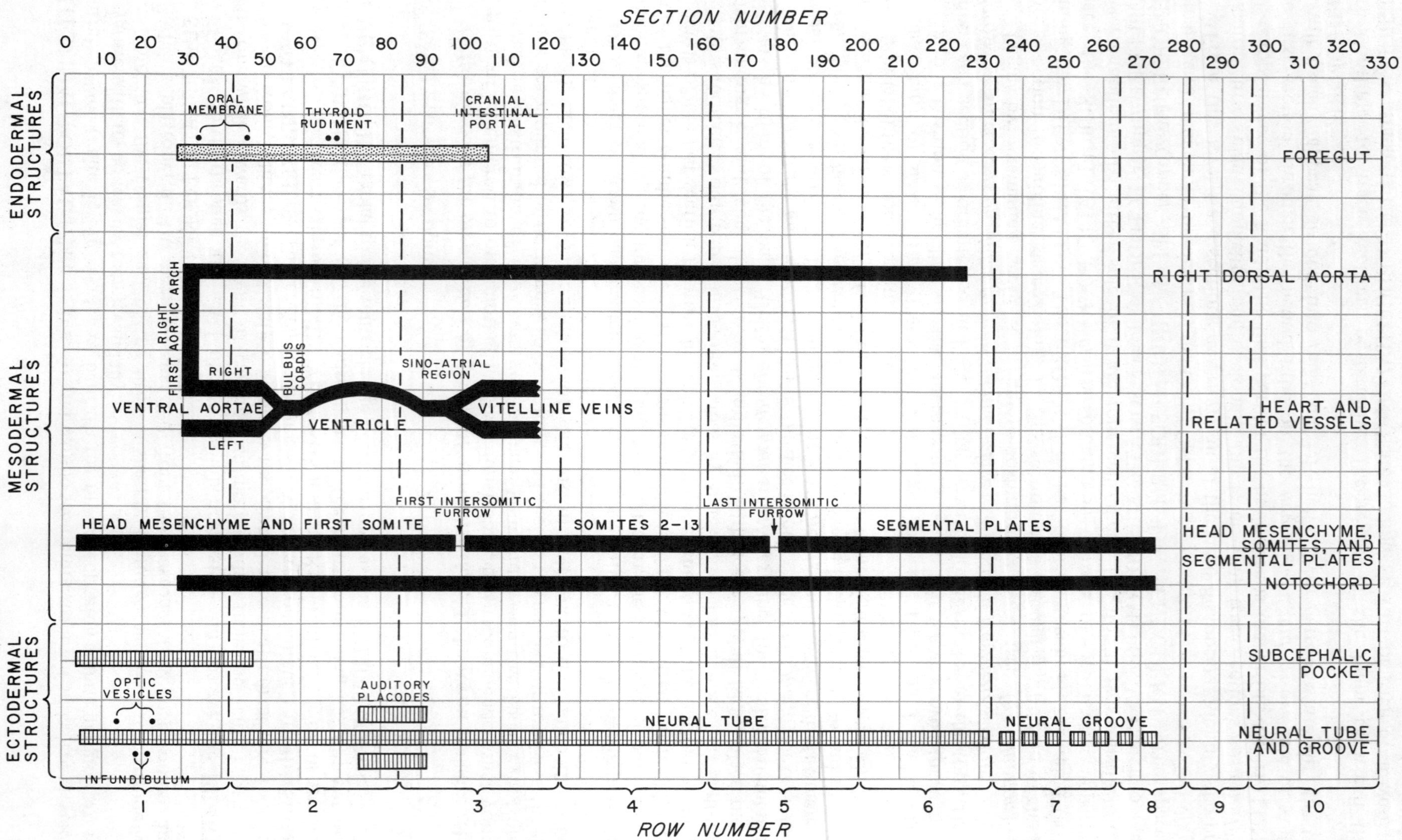

Fig. C. Graphical reconstruction of a 33-hour chick embryo showing relationships of parts to one another along the cranio-caudal axis.

caudal to its connection to the diencephalon and thus appear in some sections as an isolated vesicle. [Record the cranio-caudal extent of the optic vesicles by dots above the *neural tube and groove* line. If your embryo is not sectioned symmetrically, record these boundaries for one side only.] As the optic vesicles begin to fade out, note an evagination from the floor of the diencephalon, the INFUNDIBULUM. [Record the cranio-caudal extent of the infundibulum by dots to the left of the *neural tube and groove* line.] Identify the thin SKIN ECTODERM enclosing the head, and the HEAD MESENCHYME between the skin ectoderm and brain wall.

A few sections behind the infundibulum a small mass of cells appears beneath the neural tube, closely applied to its surface. This is the cranial tip of the NOTOCHORD. [Record the first section through the notochord on the *notochord* line.] Continue to trace neural tube and notochord caudad, noting that the notochord becomes more distinct and separated from the neural tube. The level of the neural tube which is oval- or pear-shaped is the MESENCEPHALON. It appears as the optic vesicles fade out, just caudal to the infundibulum. In more posterior sections the neural tube usually becomes smaller. In this general region the METENCEPHALON can be identified as that portion of the neural tube which lies above the bulbus cordis (Plate 2, Fig. 4). The caudal end of the SUBCEPHALIC POCKET may also be identified at about this level. [Record the cranio-caudal extent of the subcephalic pocket by a blue bar on the appropriate line under *ectodermal structures*.] In more posterior sections the MYELENCEPHALON can be identified as that portion of the neural tube which lies above the ventricle and sino-atrial region (Plate 3, Figs. 1, 2). Alongside part of the myelencephalon the skin ectoderm is thickened as the AUDITORY (OTIC) PLACODES, the rudiments of the INNER EARS. [Record the cranio-caudal extent of the auditory placodes by blue bars above and below the *neural tube and groove* line.] Formation of the auditory placodes is induced by the myelencephalon and adjacent head mesenchyme.

Focus on the dorsal surface of the myelencephalon just cranial to or just caudal to the auditory placodes. You may be able to identify a straggling line of cells wedged between the skin ectoderm and myelencephalon (HU, Fig. 108d; M, Fig. 100; P, Fig. 63D). These are NEURAL CREST CELLS, which are derived from *ectoderm.* Both neural crest cells and mesoderm contribute to the HEAD MESENCHYME.

Note the change in shape of the neural tube as you continue caudally, indicating that you have reached the level of the SPINAL CORD. The lateral walls of the spinal cord are much thicker than its dorsal and ventral walls. Recall that in whole mounts only the thick lateral walls of the neural tube can be readily seen. The dorsal wall is designated as the ROOF PLATE, the ventral as the FLOOR PLATE. The SKIN ECTODERM can be identified overlying the roof plate. In more posterior sections the neural tube may appear to contain no cavity. If so, this is probably due to shrinkage during section preparation. As sections are traced posteriorly the roof plate disappears; this is the beginning of the NEURAL GROOVE (Plate 3, Fig. 4). [Record the cranio-caudal extent of the neural tube by a blue bar on the *neural tube and groove* line.] The neural groove is flanked by the NEURAL FOLDS. Continue following the neural groove caudad and note the progressive enlargement and flattening of the notochord. The notochord then becomes indistinct, appearing to fuse with the floor of the neural groove. [Record the cranio-caudal extent of the notochord by a red bar on the *notochord* line.] The neural folds are widely separated at this level (see Plate 1, Fig. 1). Slightly more caudally a large mass of cells, which extends across the midline, gradually appears in place of the notochord. You are now near the caudal end of the neural groove in the region of the PRIMITIVE STREAK. [Record the cranio-caudal extent of the neural groove by a broken blue bar on the *neural tube and groove* line.] In more posterior sections identify the V-shaped PRIMITIVE GROOVE, which is flanked by the PRIMITIVE RIDGES (Plate 3, Fig. 5). The primitive ridges are directly continuous with the neural folds, so the transition between the U-shaped neural groove and the V-shaped primitive groove is gradual. In more posterior sections the primitive ridges and groove disappear, indicating that you are now caudal to the primitive streak.

Return in your sections to the level of the optic vesicles and again trace sections posteriorly. A small endoderm-lined cavity, the FOREGUT, will appear just beneath the neural tube. [Record the cranial boundary of the foregut on the *foregut* line.] In more posterior sections the foregut widens and its endodermal floor contacts the ventral skin ectoderm of the head, forming the ORAL MEMBRANE. If a space exists between the two layers, it is a shrinkage space caused by section preparation. These two layers will *later* rupture to establish the MOUTH OPENING. [Record the cranio-caudal extent of the oral membrane by dots above the *foregut* line.]

As you continue to trace the foregut caudad examine its floor carefully. Identify a midline thickening, the THYROID RUDIMENT, which usually appears as a shallow depression above the ventricle (Plate 3, Fig. 1). [Record the cranio-caudal extent of the thyroid rudiment by dots above the *foregut* line.] Continue to trace the foregut caudad and note the sudden disappearance of its floor; this is the region of the CRANIAL INTESTINAL PORTAL (see H, Fig. 108d). [Record the cranio-caudal extent of the foregut by a yellow bar on the *foregut* line, and label the level of the cranial intestinal portal.] Note that more caudally the embryo progressively flattens dorso-ventrally, and that endoderm forms its lower layer (Plate 3, Fig. 3). The endoderm becomes indistinguishable at the level of the primitive streak.

Return again to the level of the rhombencephalon and identify the loosely packed cells of the HEAD MESENCHYME. Just caudal to the auditory placodes the mesoderm to either side of the neural tube forms compact masses, the SOMITES. The first somite pair is difficult to identify since it has no distinct cranial boundaries. Two to three sections posteriorly the first somite pair begins to fade out as sections cut through the first INTERSOMITIC FURROWS. [Record the approximate cranio-caudal extent of the head mesenchyme and first somite pair by a red bar on the *head mesenchyme, somites, and segmental plates* line.] Slightly more caudally the second somite pair appears and soon begins to fade out at the level of the second intersomitic furrows. Continue tracing the somites caudad and determine the approximate number of pairs present in your embryo. [Record the distance between the first and last intersomitic furrows by a red bar on the *head mesenchyme, somites, and segmental plates* line, and indicate how many somite pairs this bar represents.]

Examine in detail the structure of one of the best-formed somites under higher magnification. It is roughly triangular in shape and usually contains a small core of cells (Plate 3, Fig. 3). Note the close relationship of the somites and notochord to the neural tube. Experiments have shown that the characteristic shape of the neural tube (that is, thick lateral walls and thin roof and floor plates) depends on this relationship. At this level it may be possible to observe NEURAL CREST CELLS lying between the neural tube, skin ectoderm, and somites.

Just lateral to each somite is a slender area of mesodermal cells, the NEPHROTOME (INTERMEDIATE MESODERM). At several somite levels the nephrotome on each side buds off a mass of cells called the PRONEPHRIC CORD (Plate 3, Fig. 3). Collectively these cords constitute the paired rudimentary PRONEPHRIC KIDNEYS. The pronephric cords on each side *later* elongate and fuse to form a MESONEPHRIC (PRONEPHRIC) DUCT RUDIMENT. The mesoderm lateral to the nephrotome is the LATERAL PLATE. It is divided into an upper layer of SOMATIC MESODERM (adjacent to the ectoderm) and a lower layer of SPLANCHNIC MESODERM (adjacent to the endoderm). The space between the somatic and splanchnic mesoderm is the COELOM. Note the presence of VITELLINE BLOOD VESSELS in the splanchnic mesoderm.

In sections posterior to the last pair of intersomitic furrows identify the SEGMENTAL PLATES. As you trace them caudad observe that they become continuous with the more lateral mesoderm (Plate 3, Fig. 4), and that they become indistinguishable more caudally at the level of the primitive streak. [Record the cranio-caudal extent of the segmental plates by a red bar on the *head mesenchyme, somites, and segmental plates* line.]

There remains for consideration only the circulatory system which develops from mesoderm (see A, Fig. 544; HU, Figs. 105, 106, 108b-e; M, Fig. 94; P, Figs. 57, 60D, 63B-E). Return to the level of the diencephalon and again trace sections posteriorly. At about the level where the cranial end of the foregut appears a white space should be visible on each side of the foregut. These spaces are the FIRST AORTIC ARCHES (Plate 2, Fig. 2). [Record the cranio-caudal extent of the right first aortic arch by a red bar interconnecting the sixth and ninth centimeter lines under *mesodermal structures*.] Each is continuous with a small VENTRAL AORTA, lying below the foregut, and with a larger DORSAL AORTA, lying above the foregut. Since the first aortic arch is often very small it may not be readily seen in your sections.

Trace sections posteriorly watching all aortae. The two ventral aortae may be difficult to follow since they are just forming. Note that they merge with the BULBUS CORDIS caudal to the oral membrane. [Record the cranio-caudal extent of the ventral aortae by red bars on the fifth and sixth centimeter lines under *mesodermal structures.* Connect the right ventral aorta with the right first aortic arch.] The bulbus cordis lies in the midline and has a small diameter. It consists of an inner, very thin lining of ENDOCARDIUM, continuous with the endothelium of the ventral aortae, reinforced by an outer thick layer, the EPI-MYOCARDIUM. While tracing sections posteriorly notice that the heart usually bends from the midline somewhat towards the *embryo's right,* the *apparent left* (Plate 3, Fig. 1). This is the region of the VENTRICLE. Even if the ventricle does not yet bend to the right in your embryo, it can be readily identified since its endocardium and epi-myocardium are widely separated. Note that the ventricle is suspended within the coelom by a bridge of mesoderm, the DORSAL MESOCARDIUM. The portion of the coelom surrounding the heart is the PERICARDIAL CAVITY. In more posterior sections the endocardium and epi-myocardium are no longer widely separated and the heart lies in the midline. This is the beginning of the SINO-ATRIAL REGION. The sino-atrial region is incompletely formed caudally, consisting of two rudiments, one on either side of the cranial intestinal portal. As you continue tracing sections posteriorly there is a gradual transition between the sino-atrial region and VITELLINE VEINS. The vitelline veins shift laterally and consist only of endothelium. [Record the approximate cranio-caudal extent of the bulbus cordis, ventricle, sino-atrial region, and vitelline veins.]

Shift your attention to the dorsal aortae and trace them caudad. Throughout most of their extent the dorsal aortae lie just below the nephrotomes and somites (Plate 3, Fig. 3). Their endothelium is so thin that it may not be visible in some sections. [Record the cranio-caudal extent of the right dorsal aorta by a red bar on the *right dorsal aorta* line. Connect the right dorsal aorta with the right first aortic arch.]

3. **Serial sagittal sections. Aids:** Plate 3, Fig. 6; A, Fig. 545; M, Fig. 93.

Careful study of serial sagittal sections can help you immeasurably to understand relationships among developing parts of early chick embryos. These sections are cut lengthwise in the dorso-ventral plane. First examine your slide with the naked eye. Only the darkest sections are cut through the body of the embryo. Position your slide on the microscope stage so that when viewed through the microscope the neural tube is uppermost (Plate 3, Fig. 6). Trace sections until you find one closely resembling Plate 3, Fig. 6. In midsagittal sections identify the various levels of the NEURAL TUBE. Caudal to the neural tube identify the NEURAL GROOVE. The NOTOCHORD lies beneath the neural tube and groove, but only part of the notochord's length will be cut through in any one section. Determine the cranio-caudal extent of the notochord. Caudally it merges with the PRIMITIVE STREAK. Its cranial end lies just caudal to the INFUNDIBULUM. Identify the endoderm-lined FOREGUT, the double-layered ORAL MEMBRANE, and the CRANIAL INTESTINAL PORTAL. Note that the head is separated from the PROAMNION by the SUBCEPHALIC POCKET.

Beneath the foregut is the portion of the COELOM containing the heart, the PERICARDIAL CAVITY. Identify the BULBUS CORDIS, VENTRICLE, and SINO-ATRIAL REGION. To identify these structures it may be necessary to trace sections to either side of the midline. By carefully tracing sections you may also be able to identify the DORSAL AORTAE, above the foregut, the VENTRAL AORTAE, below the foregut, and the FIRST AORTIC ARCHES.

In sections to either side of the midline note that as the PROSENCEPHALON fades out, the OPTIC VESICLES remain for several more sections. Also note that as the MYELENCEPHALON begins to fade out, the skin ectoderm overlying it thickens as the AUDITORY PLACODES. In sections to either side of the neural tube identify the HEAD MESENCHYME, SOMITES, INTERSOMITIC FURROWS, and SEGMENTAL PLATES.

4. Summary of germ layer derivatives.

Ectoderm	*Mesoderm*	*Endoderm*
auditory placodes	blood islands	foregut
infundibulum	blood vessels (endothelium)	thyroid rudiment
neural crest cells	dorsal mesocardium	
neural folds and groove	heart	
neural tube	lateral plates	
optic vesicles	nephrotomes	
	notochord	
	primitive blood cells	
	pronephric cords	
	segmental plates and somites	

Ectoderm and Mesoderm
head mesenchyme

Ectoderm and Endoderm
oral membrane
proamnion

C. STRUCTURE AND FUNCTION OF THE REPRODUCTIVE SYSTEM OF THE ADULT HEN

Adult hens possess a functional reproductive system (OVARY and OVIDUCT) only on the *left* side of the body. The fully developed ovary (Fig. D) consists of a mass of protruding FOLLICLES. It usually contains five to six larger follicles and many smaller ones. The smallest follicles are called PRIMARY FOLLICLES. The larger ones are rapidly GROWING FOLLICLES, the largest of which is about to rupture within an area called the STIGMA. The rupture of the follicle and the release of its contained ovum constitutes OVULATION. Each ovulation is apparently triggered by a sudden release of LUTEINIZING HORMONE (LH) from the ANTERIOR PITUITARY GLAND (ADENOHYPOPHYSIS). The ovary illustrated also contains a COLLAPSED FOLLICLE from which an ovum has just escaped (Fig. D: ovum after ovulation) through the ruptured stigma.

The fully developed oviduct is a highly differentiated organ and is subdivided into five distinct regions: INFUNDIBULUM, MAGNUM, ISTHMUS, SHELL GLAND, and VAGINA. The first subdivision of the oviduct is the INFUNDIBULUM. Cranially it expands and forms a delicate, funnel-shaped structure with processes called FIMBRIA along its margin. The ovum enters the opening of the infundibulum, the OSTIUM, within 15 minutes after ovulation. Within the infundibulum, or perhaps within the next subdivision of the oviduct, the magnum, the OUTER VITELLINE MEMBRANE is added to the ovum. The ovum remains in the infundibulum for about 15 minutes before it is carried by peristalsis to the magnum.

There is no clear line of demarcation between the infundibulum and the longest subdivision of the oviduct, the MAGNUM. In the magnum primarily ALBUMEN (EGG WHITE) is added to the ovum. The albumen is a mixture of proteins, most of which protect the embryo from infection. It takes about 3 hours for the ovum to traverse the magnum.

The next region of the oviduct, the ISTHMUS, secretes the INNER and OUTER SHELL MEMBRANES. The ovum remains within the isthmus for about 1 hour and is then carried to the SHELL GLAND. The primary function of the shell gland is formation of the calcified SHELL, a process which requires about 20 hours. The ovum is retained in the shell gland during this period by a sphincter muscle which separates the shell gland from the last region of the oviduct, the VAGINA. During laying of the bird's egg, or OVIPOSITION, the sphincter muscle is relaxed and the egg is rapidly forced through the vagina to the outside by contractions of the shell gland and abdominal musculature.

D. OOGENESIS, FERTILIZATION, AND CLEAVAGE

Development of the ovum (OOGENESIS) begins in the *embryonic* left ovary with formation of small cells called OOGONIA. These cells rapidly increase in number by mitotic division. *Before*

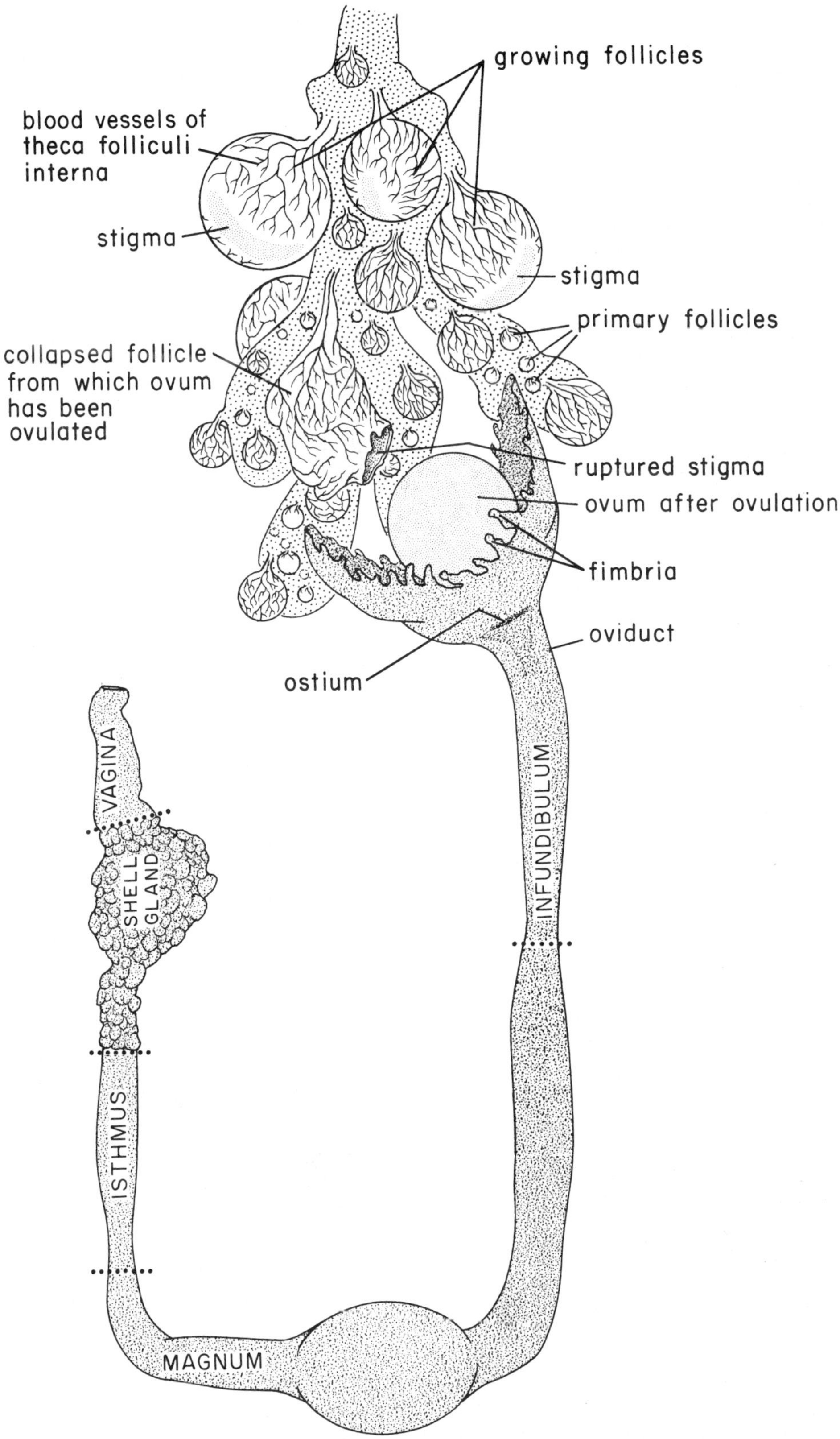

Fig. D. The functional reproductive system of the mature hen. The ovary is the lobulated structure at the top of the drawing; below it the subdivisions of the oviduct are illustrated. The magnum is greatly distended by the passage of an ovum.

the embryo hatches, oogonia stop dividing mitotically and enlarge slightly to form PRIMARY OOCYTES, 1-2 mm in diameter. Early in the primary oocyte stage each chromosome and its contained DNA are replicated. Shortly *after* hatching each primary oocyte becomes enclosed by a single layer of cells, called FOLLICLE CELLS, derived from the surface layer of the ovary. The composite structure composed of a primary oocyte enclosed by a layer of follicle cells is called a PRIMARY FOLLICLE. These structures cause the surface of the ovary to bulge outward (Fig. D). Starting at puberty, and at intervals thereafter, a group of primary follicles begins to enlarge rapidly due to the accumulation of YOLK by the primary oocytes. Yolk accumulation is primarily stimulated by FOLLICLE STIMULATING HORMONE (FSH) from the ANTERIOR PITUITARY GLAND. Yolk components are synthesized by the hen's liver, transported to the blood vessels in the connective tissue surrounding the follicle, the THECA FOLLICULI INTERNA, and then are transported to the primary oocyte via the follicle cells. As a result of this rapid accumulation of yolk, each primary follicle increases in diameter from about 5 mm to 35 mm. As the yolk accumulates, most of the cytoplasm of the oocyte is crowded off to one side as a small circular disc, the BLASTODISC, approximately 3.5 mm in diameter. The blastodisc contains the enlarged nucleus of the primary oocyte, the so-called GERMINAL VESICLE. Meanwhile, during the rapid growth phase, the INNER VITELLINE MEMBRANE forms between the primary oocyte and adjacent follicle cells.

Shortly before ovulation occurs the primary oocyte undergoes the FIRST MEIOTIC DIVISION. The meiotic spindle involved in this division forms within the blastodisc, and as the division is completed a tiny FIRST POLAR BODY is pinched off between the inner vitelline membrane and the large SECONDARY OOCYTE. The latter contains most of the yolk plus the blastodisc. Soon after the first meiotic division occurs, the spindle for the SECOND MEIOTIC DIVISION forms within the blastodisc, and the second meiotic division proceeds to the metaphase stage and then stops. At this point the ovum (although still a secondary oocyte arrested in the metaphase stage of the second meiotic division) is sufficiently mature to be fertilized and is therefore ready for ovulation. One ovum will be ovulated daily over a period of 2 or more days constituting a CLUTCH of ova. Ovulation will then cease for at least 1 day, after which it will resume in a new clutch of ova.

Following ovulation the ovum enters the infundibulum of the oviduct, the site of FERTILIZATION. For fertilization to occur sperm must penetrate at least the *inner* vitelline membrane to reach the surface of the blastodisc. It is unknown whether sperm can also penetrate the *outer* vitelline membrane, but it is known that following addition of albumen in the magnum, fertilization can no longer occur. Recall that at ovulation the ovum is a secondary oocyte arrested in the metaphase stage of the second meiotic division. As a successful sperm contacts and penetrates the blastodisc, the second meiotic division is completed, resulting in formation of a SECOND POLAR BODY and a MATURE OVUM containing the FEMALE PRONUCLEUS. Within the cytoplasm of the blastodisc the sperm nucleus enlarges to form the MALE PRONUCLEUS, and the two pronuclei closely approach one another. If the ovum is not fertilized it remains as a secondary oocyte arrested in the metaphase stage of the second meiotic division, and the cytoplasmic cap on the yolk remains as the undeveloped blastodisc.

Following fertilization the blastodisc initiates CLEAVAGE (that is, is subdivided into cells called BLASTOMERES by mitosis) and is subsequently designated as the BLASTODERM. Cleavage occurs only within the cytoplasm (the yolk is not cleaved), and the CLEAVAGE FURROWS neither extend all the way to the periphery of the cytoplasm nor do they cut entirely through its thickness. Cleavage in the hen's ovum is therefore classified as PARTIAL (MEROBLASTIC); it is also classified as DISCOIDAL because it is restricted to the circular disc of cytoplasm.

Cleavage normally begins as the fertilized ovum enters the isthmus of the oviduct, approximately 3½ hours after ovulation. The first cleavage furrow (Fig. Ea) extends across the central portion of the cytoplasm, partially separating it into two blastomeres. The second cleavage furrow forms at approximately right angles to the first and about 15 minutes later (Fig. Eb). The result is four partially separated blastomeres. While the ovum is still within the isthmus, the blastoderm usually undergoes the third cleavage (Fig. Ec). Two third-cleavage

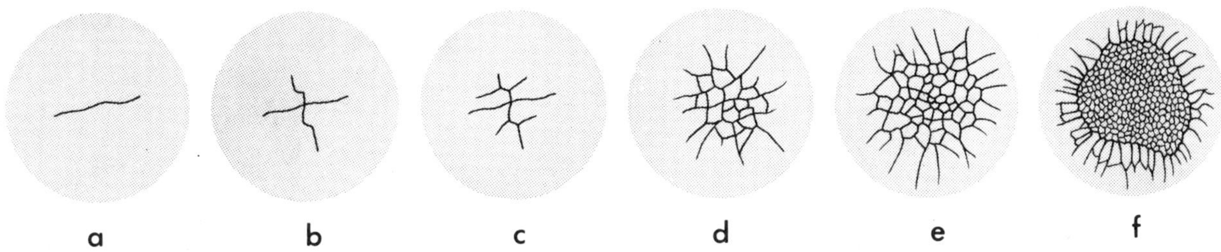

Fig. E. Early cleavage stages of the chick blastoderm removed from the surface of the yolk and viewed from the upper surface. (a) 2-cell stage showing first cleavage furrow; (b) 4-cell stage showing first and second cleavage furrows; (c) 8-cell stage showing two third cleavage furrows in addition to the first and second; (d) approximate 32-cell stage; (e) and (f) progressively more advanced cleavage stages.

furrows form at approximately right angles to the second cleavage furrow and parallel to the first, forming eight partially separated blastomeres. Later cleavages usually take place in the shell gland (Figs. Ed-Ef). The size and shape of the blastomeres become variable by the 32-cell stage (Fig. Ed). At this time it is possible to distinguish CENTRAL BLASTOMERES, which appear to be completely bounded in surface view, from MARGINAL BLASTOMERES, which still are incompletely separated from one another in surface view. The central blastomeres are separated from the yolk by a cavity, the SUBGERMINAL CAVITY, whereas the marginal blastomeres contact the yolk. More advanced cleavage stages (Figs. Ee, Ef) possess increased numbers of smaller blastomeres, some formed by division of marginal blastomeres. By the stage illustrated by Fig. Ef the central portion of the blastoderm will probably be more than one cell thick. This is because cleavage furrows also form horizontally in this region, separating outer cells from inner cells.

E. FORMATION OF THE HYPOBLAST

While the developing ovum is still present in the shell gland, and after formation of the subgerminal cavity beneath the central cells, two distinct regions can be readily distinguished in surface view: the lighter, central AREA PELLUCIDA, and the darker, peripheral AREA OPACA. Beneath the area pellucida a lower layer of large, yolky cells forms. This layer of cells is called the HYPOBLAST (Fig. Fa). Following formation of the hypoblast the surface layer of the area pellucida is called the EPIBLAST. Initially the hypoblast consists of a single layer of cells closely linked together only beneath the caudal portion of the area pellucida, whereas more cranially the hypoblast consists only of scattered cells. The exact mechanism by which the hypoblast forms is unknown. Observations suggest, however, that cells (1) separate from the area pellucida and move inward (that is, INGRESS), as well as (2) spread craniad from the caudal portion of the area pellucida (that is, from a so-called PROLIFERATION AND MIGRATION CENTER). The fertilized egg is usually laid during the stage of hypoblast formation. Further development of the egg requires incubation.

F. GASTRULATION

Our understanding of events occurring during GASTRULATION (the process of germ layer formation) has been clarified by experiments using tritiated thymidine as a cell marker. Tritiated thymidine is incorporated specifically in DNA, thus radioactively labeling cells. Specific sectors of the epiblast from labeled donor embryos are used to replace corresponding sectors of unlabeled host embryos of the same stage. Movements of labeled cells can then be followed precisely during gastrulation and their fates determined.

1. Location and migration of the prospective endoderm. Aid: Fig. F.

Labeling experiments have clearly demonstrated that PROSPECTIVE ENDODERM is located in an oval-shaped region of the epiblast of prestreak stages, and around and within the

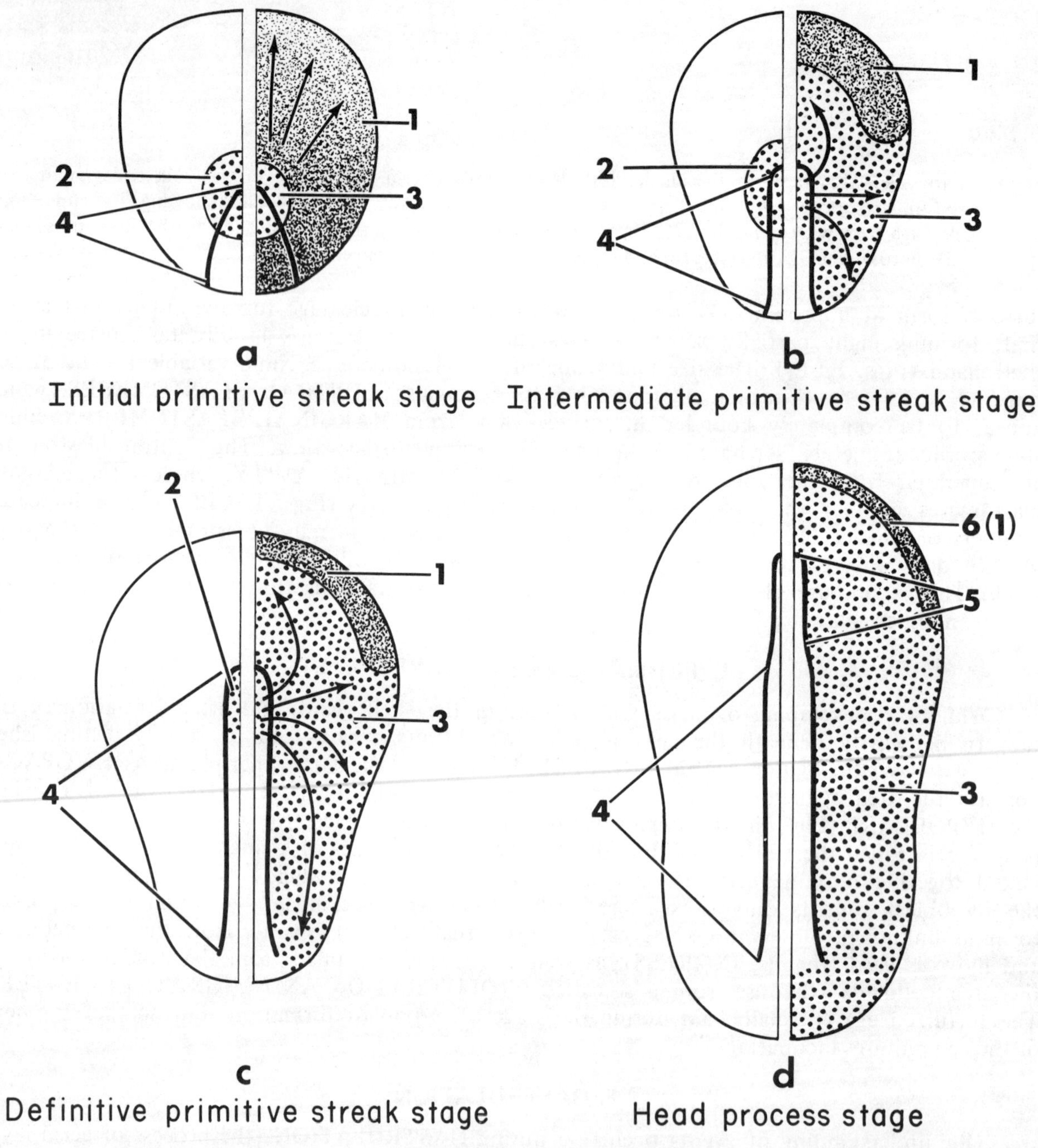

Fig. F. Diagrams of the area pellucida showing formation of the hypoblast and involuted endoderm. The epiblast is shown on the left side of each diagram, but not on the right side. Arrows indicate the direction of displacement of the hypoblast in Fig. Fa, and directions of migration of the involuted endodermal cells in Figs. Fb, Fc.

1. Hypoblast
2. Prospective endoderm
3. Endoderm
4. Primitive streak
5. Head process
6. Germ cell crescent

K (Fig. Fa). The primitive streak is a midline
e streak stage prospective endoderm is migrating
e streak. The inward migration of epiblast cells
VOLUTION. The involuted ENDODERM enters
e hypoblast cranio-laterally (Fig. Fa). Involution
e hypoblast continue rapidly in the intermediate
ive primitive streak stage the only prospective
cranial end of the primitive streak (Fig. Fc).
ength at this stage, extending about two-thirds
he primitive streak is responsible for the change
(Fig. Fa) to pear-shaped (Fig. Fc). By the head process stage the epiblast no longer contains any prospective endoderm since it has all undergone involution into the interior (Fig. Fd). The displaced hypoblast becomes condensed into a crescent-shaped area at the cranio-lateral margin of the area pellucida. This area is designated as the GERM CELL CRESCENT since it contains the PRIMORDIAL GERM CELLS, presumably of hypoblast origin. The primordial germ cells *later* undergo extensive migration and enter the embryonic gonads where they form OOGONIA in the female (see page 12), and SPERMATOGONIA in the male.

2. Location and migration of the prospective mesoderm. Aid: Fig. G.

By the intermediate primitive streak stage (Fig. Ga) involution of the PROSPECTIVE EXTRA-EMBRYONIC MESODERM, through the caudal half of the primitive streak, and the cranio-lateral internal spreading of the involuted EXTRA-EMBRYONIC MESODERM have begun. This first mesoderm to involute is designated as extra-embryonic since it does not contribute to the body of the embryo. Located around and within the cranial half of the intermediate primitive streak are the PROSPECTIVE HEAD MESENCHYME and PROSPECTIVE NOTOCHORD. These subdivisions of PROSPECTIVE EMBRYONIC MESODERM have not yet started their involution at this stage.

At the definitive primitive streak stage (Fig. Gb) involution of the PROSPECTIVE EXTRA-EMBRYONIC MESODERM through the caudal half of the primitive streak continues, and the internal spreading of the involuted EXTRA-EMBRYONIC MESODERM is now caudo-lateral as well as cranio-lateral. All of the PROSPECTIVE NOTOCHORD now occupies the cranial end of the primitive streak. Just caudal to the prospective notochord PROSPECTIVE HEAD MESENCHYME is undergoing involution, and the involuted HEAD MESENCHYME is spreading internally. Likewise, just caudal to the prospective head mesenchyme PROSPECTIVE HEART MESODERM is involuting, and the involuted HEART MESODERM is spreading internally. At this stage PROSPECTIVE SEGMENTAL PLATE MESODERM and PROSPECTIVE LATERAL PLATE MESODERM are located in the epiblast, just lateral to the cranial half of the primitive streak. These subdivisions of prospective embryonic mesoderm have not yet started their involution.

At the head process stage (Fig. Gc) involution of the PROSPECTIVE EXTRA-EMBRYONIC MESODERM and the internal spreading of the involuted EXTRA-EMBRYONIC MESODERM are still under way. The PROSPECTIVE HEAD MESENCHYME and PROSPECTIVE HEART MESODERM have now completed their involution. The PROSPECTIVE NOTOCHORD has already started involuting through the cranial end of the primitive streak by this stage, forming the so-called HEAD PROCESS. The head process is a mesodermal tongue of cells which subsequently forms the cranial part of the NOTOCHORD. Just caudal to the prospective notochord PROSPECTIVE SEGMENTAL PLATE MESODERM and PROSPECTIVE LATERAL PLATE MESODERM are involuting, and the involuted SEGMENTAL PLATE MESODERM and LATERAL PLATE MESODERM are spreading internally.

3. Summary of epiblast involution during gastrulation.

Initial Primitive Streak Stage

Prospective endoderm is involuting through the cranial end of the primitive streak.

Intermediate Primitive Streak Stage

Prospective endoderm continues to involute through the cranial end of the primitive streak.

Prospective extra-embryonic mesoderm is involuting through the caudal half of the primitive streak.

Definitive Primitive Streak Stage

Prospective endoderm continues to involute through the cranial end of the primitive streak.

Prospective head mesenchyme and *prospective heart mesoderm* are involuting through the cranial half of the primitive streak.

Prospective extra-embryonic mesoderm continues to involute through the caudal half of the primitive streak.

Head Process Stage

Involution of *prospective endoderm, prospective head mesenchyme,* and *prospective heart mesoderm* is complete.

Prospective notochord, prospective segmental plate mesoderm, and *prospective lateral plate mesoderm* are involuting through the cranial half of the primitive streak.

Prospective extra-embryonic mesoderm continues to involute through the caudal half of the primitive streak.

4. Location and changes of the prospective ectoderm. Aid: Fig. G.

The region of the epiblast which does not undergo involution remains on the surface as the ECTODERM. As involution of prospective endoderm and mesoderm occurs, the prospective ectoderm expands, spreading towards the primitive streak (that is, it undergoes EPIBOLY) to replace those areas of the epiblast which have involuted. At the head process stage (Fig. Gc) the ectoderm overlying the head mesenchyme, head process, and segmental plate mesoderm has thickened forming the NEURAL PLATE, the rudiment of the central nervous system. Neural plate formation is induced primarily by the mesoderm lying beneath it. The approximate locations of the PROSPECTIVE PROSENCEPHALON, MESENCEPHALON, RHOMBENCEPHALON, and SPINAL CORD are indicated in Fig. Gc. Note that at this stage mesoderm has not yet spread into the cranio-lateral portion of the area pellucida. Spread of mesoderm into this area, the PROAMNION is considerably delayed.

5. Hypothetical location of epiblast areas prior to gastrulation. Aid: Fig. H.

Cell-marking experiments have demonstrated the fates of different areas of the epiblast during gastrulation. A PROSPECTIVE FATE MAP is a schematic representation of the location of these areas prior to the onset of gastrulation. Note that prospective extra-embryonic meso-

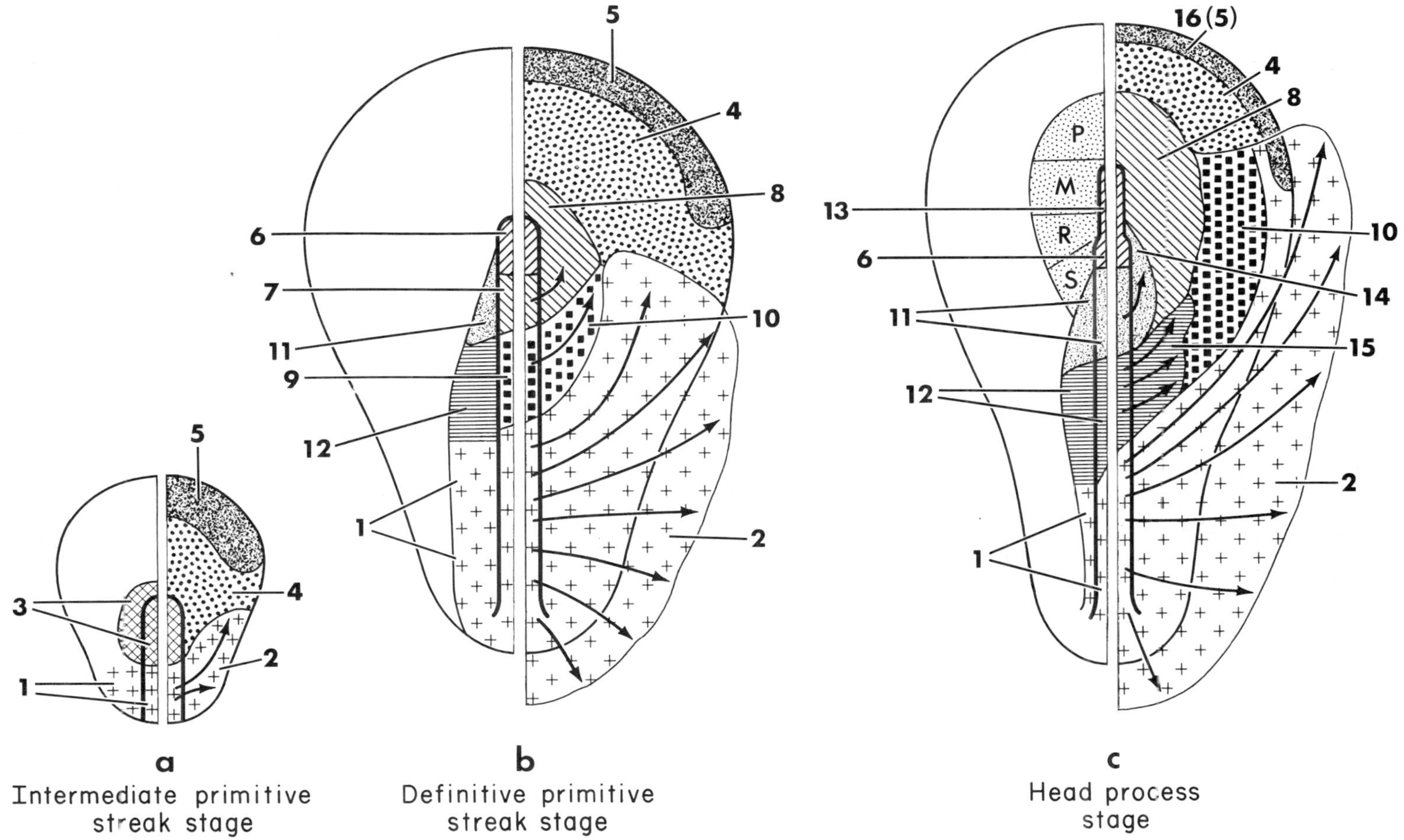

Fig. G. Diagrams of the area pellucida showing formation of the subdivisions of the involuted mesoderm. The epiblast is shown on the left side of each diagram, but not on the right side. Arrows indicate the directions of migration of the involuted mesodermal cells. The boundaries of the primitive streak and head process are indicated by a heavy black line. P, M, R, and S represent, respectively, prospective prosencephalon, mesencephalon, rhombencephalon, and spinal cord of the left half of the neural plate.

1. Prospective extra-embryonic mesoderm
2. Extra-embryonic mesoderm
3. Prospective head mesenchyme, prospective notochord
4. Endoderm
5. Hypoblast
6. Prospective notochord
7. Prospective head mesenchyme
8. Head mesenchyme
9. Prospective heart mesoderm
10. Heart mesoderm
11. Prospective segmental plate mesoderm
12. Prospective lateral plate mesoderm
13. Head process
14. Segmental plate mesoderm
15. Lateral plate mesoderm
16. Germ cell crescent

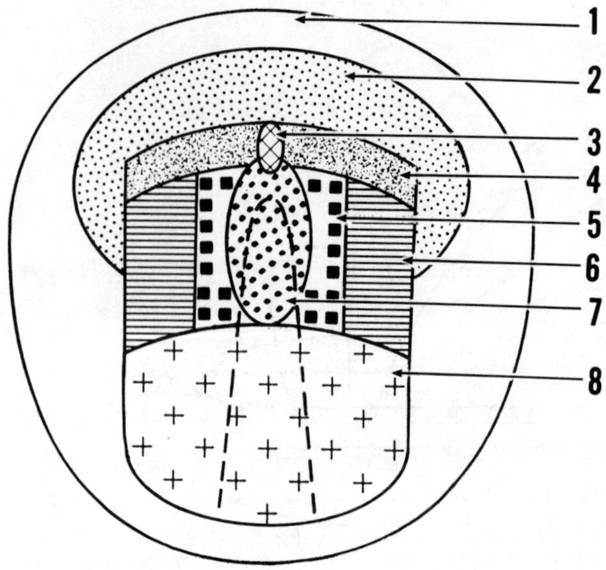

Fig. H. Prospective fate map of the chick epiblast. The approximate site of primitive streak formation is indicated by dashed lines.

1. Prospective ectoderm
2. Prospective neural plate
3. Prospective head mesenchyme, prospective notochord
4. Prospective segmental plate mesoderm
5. Prospective heart mesoderm
6. Prospective lateral plate mesoderm
7. Prospective endoderm
8. Prospective extra-embryonic mesoderm

derm is positioned at the level of formation of the caudal half of the primitive streak, through which it later involutes (Fig. H). All remaining subdivisions of the prospective mesoderm and the prospective endoderm are positioned at the level of formation of the cranial half of the primitive streak, through which they later involute. Prospective ectoderm is located peripheral to the regions of the epiblast which undergo involution.

G. 18-HOUR CHICK EMBRYOS

A semidiagrammatic surface and cross-sectional view of the cranial half of the blastoderm and definitive primitive streak is illustrated by Fig. I. The cranial end of the primitive streak is thickened as the PRIMITIVE KNOT (HENSEN'S NODE). It partially surrounds a depression, the PRIMITIVE PIT, which is directly continuous caudally with the PRIMITIVE GROOVE. Lateral to the primitive groove are the PRIMITIVE RIDGES. Note that cells of the epiblast are tightly apposed to one another forming an epithelium. Within the primitive groove the outer ends of the epiblast cells become narrowed and the inner ends widen forming the so-called BOTTLE (FLASK) CELLS. As the narrow outer connections to neighboring epiblast cells are lost, FILOPODIA form at the broad inner ends of the bottle cells. The involuted cells then migrate away from the primitive streak, apparently through the action of their filopodia. The involuted mesoderm consists of loosely packed cells (that is, mesenchyme) rather than an epithelium.

Examine whole mounts of embryos resembling Plate 4, Figs. 1-3. Identify all parts of the PRIMITIVE STREAK and the HEAD PROCESS when present. Although not visible in whole mounts, an ectodermal thickening, the NEURAL PLATE, overlies the head process and adjacent involuted mesoderm. Identify the AREA PELLUCIDA and AREA OPACA. Visualize and sketch a diagram of a midsagittal section of an embryo at the head process stage (see A, Fig. 525; P, Figs. 33B, 39F). Label the following structures: skin ectoderm, neural plate, head process, endoderm, primitive knot, primitive pit, and primitive groove.

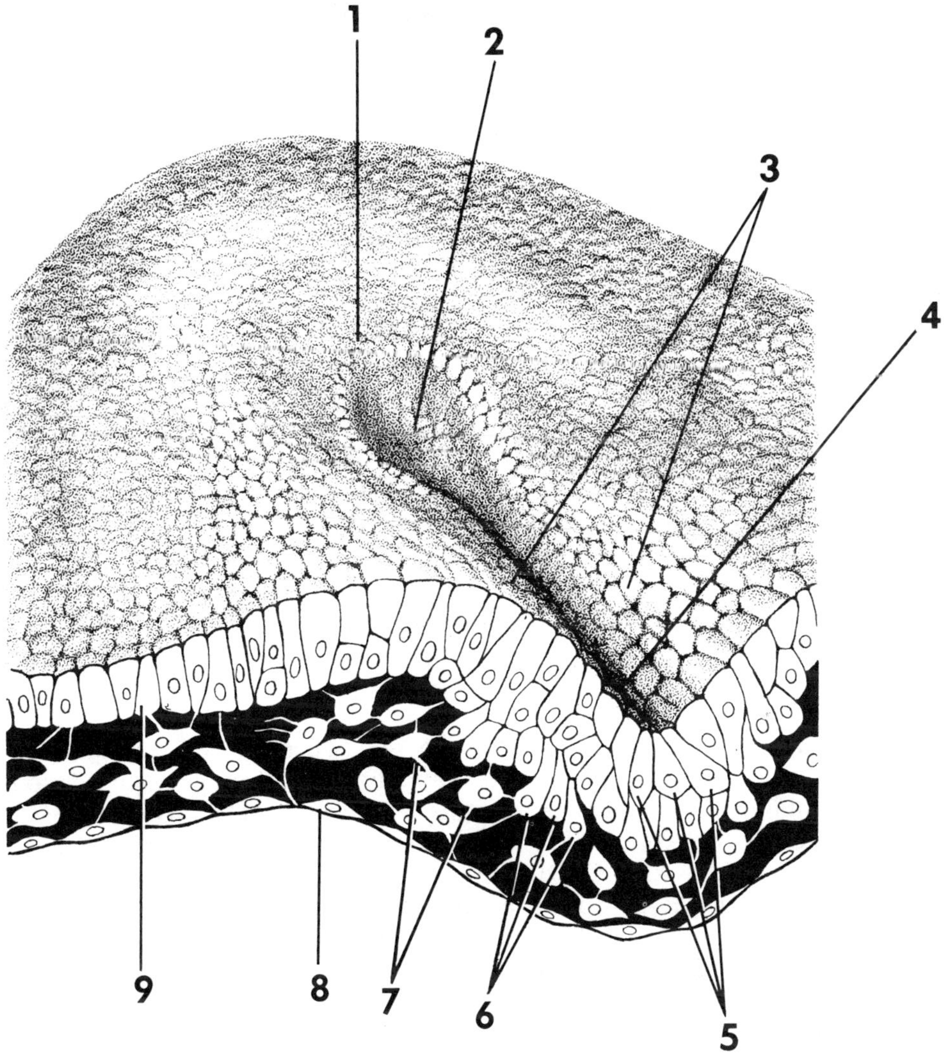

Fig. I. Three-dimensional diagram of the cranial half of a blastoderm cut transversely at the definitive primitive streak stage.

1. Primitive knot
2. Primitive pit
3. Primitive ridges
4. Primitive groove
5. Involuting prospective mesodermal cells
6. Bottle cells
7. Filopodia
8. Endodermal cell
9. Epiblast cell

Examine whole mounts of embryos resembling Plate 4, Fig. 4. Note the distinct curved line just cranial to the tip of the head process. This line is the beginning of the HEAD FOLD OF THE BODY, a ventrally directed fold of ectoderm and endoderm which establishes the cranial boundary of the head. It very quickly appears to undercut the head, forming two cavities: the ectoderm-lined SUBCEPHALIC POCKET and the endoderm-lined FOREGUT. Visualize and sketch a diagram of a midsagittal section of an embryo at the early head fold stage (see H, Fig. 50). Label the following structures: ectoderm and endoderm of the head fold of the body, subcephalic pocket, foregut, neural plate, head process, primitive knot, primitive pit, and primitive groove.

H. 24-HOUR CHICK EMBRYOS

1. Whole mount. Aids: Plate 5; A, Fig. 528; HU, Figs. 103, 104; M, Figs. 85, 86; P, Fig. 47.

Several major advances have occurred by this stage. The head process has formed the cranial part of the NOTOCHORD, and the notochord has greatly lengthened. Typically four to six pairs of SOMITES have developed from the SEGMENTAL PLATES. Although the ectoderm overlying the region of the somites and segmental plates is thickened as the NEURAL PLATE, it cannot be readily seen in whole mounts. Cranial to the neural plate distinct NEURAL FOLDS are present. These neural folds are located at the lateral margins of the NEURAL GROOVE. In some embryos they may approach one another and fuse, transforming the neural groove into the NEURAL TUBE. If the neural tube has formed in your embryo, note that its cranial end remains open as the CRANIAL NEUROPORE.

The head usually juts forward some distance above the PROAMNION so that the head fold of the body appears to have undercut the head over a considerable distance, thereby lengthening the SUBCEPHALIC POCKET and FOREGUT. The lateral margins of the foregut are usually quite distinct. The foregut opens caudally via the CRANIAL INTESTINAL PORTAL. The caudal end of the notochord is continuous with the PRIMITIVE STREAK. Try to identify the PRIMITIVE KNOT, PIT, GROOVE, and RIDGES. Note that the embryo is forming cranial to the primitive streak, as the latter steadily regresses (that is, moves caudad) and shortens.

2. Serial transverse sections. Aids: Plate 6; HU, Fig. 102; M, Figs. 87-91; P, Fig. 51.

Position your slide on the microscope stage so that when viewed through the microscope each section is oriented as in Plate 6, Figs. 1-4. Quickly trace through your sections until you find one closely resembling Plate 6, Fig. 4. At this level the NOTOCHORD and SEGMENTAL PLATES lie beneath the NEURAL PLATE. The ectoderm that has thickened to form the neural plate is designated as the NEURAL ECTODERM. It can be readily distinguished from the much thinner adjacent SKIN ECTODERM. In most embryos the neural plate is *not* a flat layer of cells. Typically it is sharply bent in the midline where it overlies the notochord (Plate 6, Fig. 4) Begin tracing sections *anteriorly* and observe that the regions of continuity between the neural and skin ectoderm begin to fold forming the NEURAL FOLDS. Each fold thus consists of an inner layer of neural ectoderm and an outer layer of skin ectoderm (Plate 6, Fig. 3). As the neural folds form and elevate, the neural plate is transformed into the NEURAL GROOVE. Continue tracing sections *anteriorly*. In some embryos the neural folds will approach one another and fuse forming the NEURAL TUBE. As the neural tube is traced craniad in these embryos note the appearance of the CRANIAL NEUROPORE. In other embryos fusion of neural folds has not yet occurred.

Locate a section cut through the cranial end of the head and trace sections *posteriorly*. Identify the two-layered PROAMNION and the SUBCEPHALIC POCKET. Sometimes the two layers of the proamnion are widely separated. In more posterior sections a mass of cells appears beneath the neural ectoderm. This mass is the endodermal cranial wall of the FOREGUT. As sections are traced posteriorly the foregut cavity appears, and the foregut widens rapidly. Note the sparsity of HEAD MESENCHYME at these levels. Identify the ORAL MEMBRANE, if present, and the NOTOCHORD between the foregut endoderm and neural ectoderm. At about this level note that mesoderm is wedged between ectoderm and endoderm on each side of the proamnion. This mesoderm is split into somatic and splanchnic layers which bound two enlarged portions of the COELOM (Plate 6, Fig. 1). Also note at this level that two small portions of the coelom are located beneath the foregut within the body of the embryo.

Locate the last sections through the subcephalic pocket. The skin ectoderm covering the lateral surfaces of the body now becomes continuous with the ectoderm of the somatopleure on either side. At the point of continuity between these two regions a distinct notch is formed by the somatopleural component of the LATERAL BODY FOLDS (Plate 6, Fig. 2). The lateral body folds consist of two components, somatopleure and splanchnopleure, and are direct continuations of the head fold of the body. By examining available models verify that these body folds are directly continuous. The lateral body folds delimit the body from extra-embryonic regions laterally. At about this level note that on each side the two portions of the coelom, the enlarged portion and the small portion, are continuous forming a large coelomic space (Plate 6, Fig. 2). These large spaces project towards the midline as the RUDIMENTS OF THE PERICARDIAL CAVITY. They later fuse to form a single cavity surrounding the heart.

At about this level, or in slightly more posterior sections, identify the CRANIAL INTESTINAL PORTAL, and note the distinct thickening of the splanchnic mesoderm to either side of the midline (Plate 6, Fig. 2). These thickenings are the CARDIAC PRIMORDIA. At a later stage cells emerge medially from each cardiac primordium and organize into an ENDOCARDIAL TUBE. These tubes fuse ventral to the foregut in cranio-caudal sequence, due to the action of the splanchnopleural component of the LATERAL BODY FOLDS, thus forming the lining of the heart, the ENDOCARDIUM. The remaining cells of each cardiac primordium form the EPI-MYOCARDIUM (see P, Figs. 59, 60).

As you continue to trace sections posteriorly, note that the embryo flattens dorsoventrally and that the mesoderm adjacent to the neural ectoderm progressively increases in density as you approach the level of the first pair of SOMITES. Try to identify the first INTERSOMITIC FURROWS and successive somites and furrows as you continue to trace sections posteriorly. Approximately how many pairs of somites does your embryo possess? Note that there is little, if any, distinction between the NEPHROTOME and LATERAL PLATE on each side.

The NEURAL GROOVE becomes progressively shallower as you continue to trace sections posteriorly and the notochord widens, changing in shape from round to oval. Identify the SEGMENTAL PLATES caudal to the somites. In more posterior sections the NEURAL FOLDS disappear and the NEURAL PLATE can then be identified. As you continue caudally the notochord increases in size and becomes indistinct, appearing to fuse with the neural plate. You are now at the level of the PRIMITIVE KNOT. Slightly caudal to this level you can usually identify the PRIMITIVE PIT. It is directly continuous caudally with the PRIMITIVE GROOVE. In some embryos there may be discontinuities in the primitive groove.

3. **Serial sagittal sections. Aids:** Plate 6, Figs. 5, 6; A, Fig. 529; P, Fig. 50E.

First examine your slide with the naked eye. Only the darkest sections are cut through the body of the embryo. Position your slide on the microscope stage so that when viewed through the microscope the neural tube is uppermost (Plate 6, Figs. 5, 6). Quickly trace sections until you first encounter the one showing the greatest number of SOMITES. How many somite pairs are present? Identify the HEAD MESENCHYME and SEGMENTAL PLATE. Try to identify the RUDIMENT OF THE PERICARDIAL CAVITY in this section or in sections nearer to the midline.

Select a section resembling Plate 6, Fig. 5. Identify the PROAMNION, SUBCEPHALIC POCKET, FOREGUT, CRANIAL INTESTINAL PORTAL, and ORAL MEMBRANE, if present. Is a NEURAL TUBE present in your embryo? If so, identify the CRANIAL NEUROPORE. Also identify the NEURAL GROOVE, and more caudally, the NEURAL PLATE. The NOTOCHORD lies directly beneath the neural ectoderm and can be identified in the midsagittal plane. Try to identify the PRIMITIVE KNOT, PIT, and GROOVE.

I. SUMMARY OF DEVELOPMENT OF THE YOUNG CHICK EMBRYO

The embryo develops in cranio-caudal sequence from germ layers established during gastrulation. Early developmental events following gastrulation may be summarized as follows:

1. **Regional specialization of the mesoderm.**

The involuted mesoderm quickly becomes specialized. The mesoderm in the midline of the embryo forms a compact rod of cells, the notochord. Cranially, the rest of the mesoderm remains loosely packed as the head mesenchyme. More caudally, the mesoderm lateral to the notochord forms three distinct zones. The mesoderm immediately adjacent to each side of the notochord forms a thick band of cells, the segmental plate, which subsequently subdivides to form a row of somites. Lateral to each row of somites the mesoderm forms a thin band of cells, the nephrotome. The latero-most mesoderm forms a thick plate of cells, the lateral plate, which splits into somatic and splanchnic layers bounding the coelom.

2. Neural tube formation.

The first indication of formation of the central nervous system is an ectodermal thickening, the neural plate. At the lateral margins of the neural plate the neural folds form. Each consists of an inner layer of neural ectoderm and an outer layer of skin ectoderm. As the neural folds form and elevate, a distinct neural groove forms. The neural folds progressively approach each other and fuse to form the neural tube.

3. Formation of body folds.

At first there is no indication as to how much of the blastoderm is destined to be *embryonic* and how much *extra-embryonic.* The first boundary of the embryo to become visible is the cranial boundary which arises in the following way. Just cranial to the neural plate the ectoderm and endoderm of the proamnion begin to fold forming the head fold of the body. The neural plate is rapidly expanding forward, and as it does so the head fold of the body appears to undercut the neural plate. To make this transformation clear, take two sheets of paper and superimpose them, considering the upper one the ectoderm, the lower one the endoderm. Hold the top edge of the papers firmly against a desk with your left hand; with your right hand grasp the bottom edges with your thumb beneath the sheets, and your other fingers above and pointing towards the top of the sheets. Then lift your right hand slightly above the desk and move it towards the top of the sheets. Both sheets of paper will elevate and project forward. This movement creates an internal bay in the endoderm, the foregut, and an external bay, the subcephalic pocket. These bays are turned in opposite directions, the internal one opening into the subgerminal cavity caudally via the cranial intestinal portal, and the external one opening cranially onto the surface of the blastoderm. Simultaneously the cranial boundary of the head is created.

The lateral boundaries of the embryo are formed by the action of the lateral body folds which are direct continuations of the head fold of the body. The lateral body folds differ from the head fold of the body in that they consist of somatopleure and splanchnopleure. The somatopleural component of these folds is involved in lengthening the subcephalic pocket. The splanchnopleural component is involved in lengthening the foregut, and in swinging together the bilateral heart rudiments beneath the lengthening foregut. In addition, the splanchnopleural component brings together the paired rudiments of the pericardial cavity.

J. 48-HOUR CHICK EMBRYOS

1. Whole mounts: injected and uninjected. Aids: Fig. J; Plates 7, 8; A, Fig. 559; H, Fig. 98; HU, Figs. 110, 111; M, Figs. 104-107; P, Figs. 74, 75, 78, 80.

First examine your uninjected whole mount slide with the naked eye. Superimpose this slide on the 33-hour whole mount and note the increased size of the embryo. The entire diameter of the BLASTODERM is not seen since it was trimmed during preparation of the slides. Position your slide on the microscope stage so that when viewed through the microscope the embryo is oriented as in Plate 7, and examine the slide under low magnification. You will note a change in the axis of the body at this stage. The cranial end of the body now lies on its *left* side (that is, it has undergone twisting or TORSION). In addition, the PROSENCEPHALON is bent towards the RHOMBENCEPHALON forming the CRANIAL (CEPHALIC) FLEXURE, a sharp bend in the floor of the MESENCEPHALON.

Examine the body of the embryo and note that it is more opaque than at the 33-hour stage, indicating its greater thickness. *Care must be taken not to focus the objective down into the elevated coverslip over these older embryos!* First examine the caudal half of the body and identify the thickened lateral walls of the NEURAL TUBE (SPINAL CORD level). Follow these two parallel lines forward until they curve towards the left and fade out. Then focus down slightly and you will see that, due to torsion, your view of the neural tube shifts from a dorsal view of the caudal half to a lateral view of the right side of the cranial half. Identify the following structures: TELENCEPHALON (the cranial neuropore is now closed); DIENCEPHALON; INFUNDIBULUM (often difficult to identify); OPTIC CUPS (formed by invagination of the optic vesicles); LENS VESICLES (formed by thickening and invagination of the ectoderm overlying the optic vesicles at the 33-hour stage); OPTIC FISSURES (ventral gap in the optic cups; see Plate 7, Fig. 2; A, Fig. 496); MESENCEPHALON (rounded area at the apex of the embryo); METENCEPHALON; ISTHMUS (a prominent constriction of the neural tube between mesencephalon and metencephalon); MYELENCEPHALON (note its thin ROOF PLATE); AUDITORY VESICLES or OTOCYSTS (formed by invagination of the auditory placodes); NOTOCHORD (usually best seen beneath the mesencephalon and rhombencephalon); FOREGUT; CRANIAL INTESTINAL PORTAL; HEART; CRANIAL LIVER RUDIMENT (an evagination of the floor of the foregut which extends towards the heart); SOMITES (count the number of pairs present); and SEGMENTAL PLATES.

Observe that the cranial half of the embryo appears to be veiled by some coverings which the caudal half lacks. The distinct curvature between covered and uncovered parts is the BOUNDARY OF THE AMNIOTIC FOLDS. The amniotic folds arch over the head, as continuous CRANIAL (ANTERIOR) and LATERAL AMNIOTIC FOLDS, forming two EXTRA-EMBRYONIC MEMBRANES: the outer CHORION and the inner AMNION. These two membranes cannot be readily distinguished from one another in whole mounts.

Note that the boundaries of the body are now quite distinct in the cranial half, whereas caudally they are probably only slightly indicated. In some embryos the TAIL FOLD OF THE BODY may have established the caudal boundary of the embryo. Following formation of the tail fold of the body, the caudal end of the embryo is separated from the underlying blastoderm by the SUBCAUDAL POCKET. In addition, a curved line with its concavity facing cranially may be visible just cranial to the caudal boundary of the body (Plate 7, Fig. 1). This curvature is the beginnings of the CAUDAL (POSTERIOR) INTESTINAL PORTAL. In most embryos the neural groove is no longer present, and the neural tube merges caudally with a dark mass of cells, the TAIL (END) BUD, which is derived from the cranial part of the primitive streak.

Observe three somewhat irregular white lines fanning out from the heart towards each auditory vesicle. The most cranial of these lines is the FIRST BRANCHIAL GROOVE, the next one is the SECOND BRANCHIAL GROOVE, and the last one is the THIRD BRANCHIAL GROOVE (Plate 7, Figs. 1, 2). Cranial to the first groove is a darkly stained mass of cells, the FIRST BRANCHIAL ARCH. This arch is partially split into two processes by another white line, the STOMODEUM (not to be confused with the first branchial groove). The dark area just cranial to the stomodeum is the MAXILLARY PROCESS of the first branchial arch. The dark area just caudal to the stomodeum is the MANDIBULAR PROCESS of the first branchial arch. Between the first and second grooves is the SECOND BRANCHIAL ARCH, and between the second and third grooves is the THIRD BRANCHIAL ARCH. *This series of structures causes no end of trouble in studying sections through this region and their relationships should be mastered now!*

Examine the HEART in uninjected embryos and in embryos whose circulatory system has been injected with india ink. (See P, Figs. 61, 103C-I, 104C-I, 105C-I to understand the changes that have occurred in the heart between the 33- and 48-hour stages.) At the cranial end of the heart identify the BULBUS CORDIS, which is continuous, without a distinct boundary, with the next region of the heart, the VENTRICLE. Notice that the ventricle is U-shaped and that it extends craniad and to the left, underneath the bulbus cordis, to become continuous with a broad saccular region, the ATRIUM. This is continuous, without a distinct boundary, with the SINUS VENOSUS. Observe that the heart consists of two layers: the inner ENDOCARDIUM, enclosing blood cells and india ink, and the outer EPI-MYOCARDIUM.

Using injected embryos note that the bulbus cordis is also continuous with a very narrow region, the AORTIC SAC (Plate 8, Fig. 2), which formed by fusion of the paired ventral aortae. Continuous with the aortic sac are two or three pairs of blood vessels, the AORTIC ARCHES. The FIRST AORTIC ARCH, on each side of the body, runs through the mandibular process and then curves sharply around the stomodeum. Likewise, on each side, the SECOND AORTIC ARCH runs through the second branchial arch, and the THIRD AORTIC ARCH, if present, runs through the third branchial arch. The aortic arches are continuous on each side with a DORSAL AORTA. The paired dorsal aortae are fused more caudally as the DESCENDING AORTA (Plate 8, Fig. 1). Numerous small blood vessels are continuous with the descending aorta dorsally; these are the INTERSEGMENTAL ARTERIES. Caudal to the descending aorta paired DORSAL AORTAE are again present.

Try to identify the PRECARDINAL (ANTERIOR CARDINAL) VEIN lying above each dorsal aorta in the region of the aortic arches (Fig. J; Plate 8, Fig. 2). This vessel is usually poorly injected. The precardinal vein on each side extends caudad from the PLEXUS OF HEAD BLOOD VESSELS and joins two other vessels (usually not readily visible): the POSTCARDINAL (POSTERIOR CARDINAL) VEIN, which extends further caudad, and the COMMON CARDINAL VEIN, which joins the sinus venosus.

Note, in injected embryos, the extensive plexus of VITELLINE BLOOD VESSELS (Plate 8, Fig. 1). These vessels are contained within the YOLK SAC, an EXTRA-EMBRYONIC MEMBRANE which overgrows the yolk. Identify the paired VITELLINE ARTERIES which are continuous with the caudal dorsal aortae at approximately right angles. The vitelline arteries and dorsal aortae are continuous at about the 22nd somite level. Also identify the paired VITELLINE VEINS which are continuous with the sinus venosus.

The circulatory system is well developed and blood circulates between the yolk sac and body of the embryo primarily in the following sequence: vitelline veins → sinus venosus → atrium → ventricle → bulbus cordis → aortic sac → aortic arches → cranial dorsal aortae → descending aorta → caudal dorsal aortae → vitelline arteries → plexus of vitelline blood vessels → vitelline veins. Some of the blood in the cranial dorsal aortae supplies the plexus of head blood vessels via small branches. The precardinal veins drain this plexus, returning blood to the sinus venosus via the common cardinal veins. Some of the blood in the descending aorta enters the intersegmental

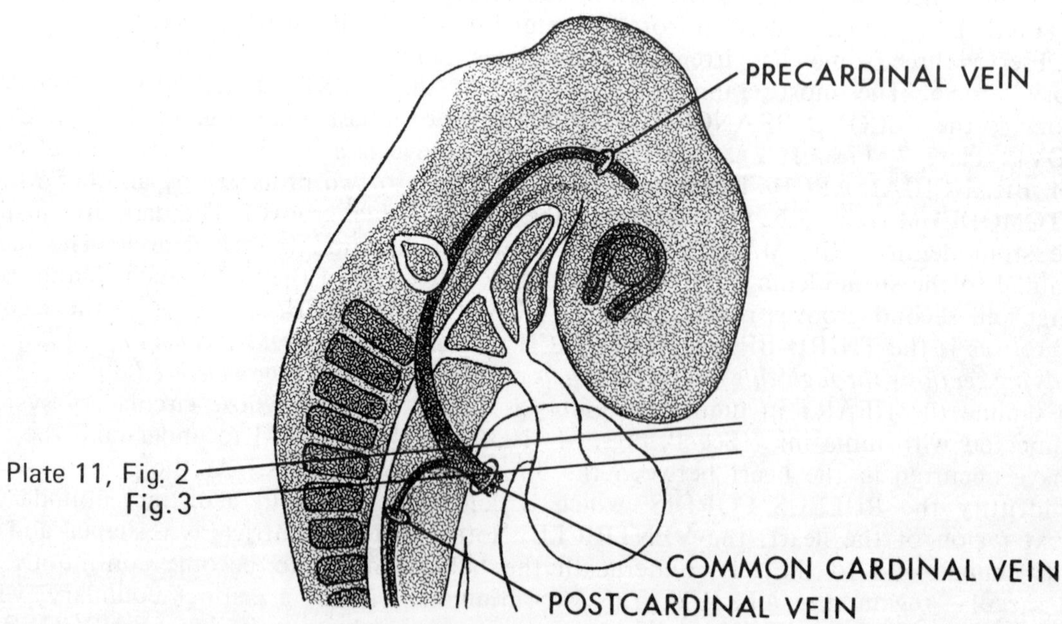

Fig. J. Semidiagrammatic drawing of the right side of the cranial half of a 48-hour chick embryo.

arteries. This blood is returned to the heart in the following sequence: postcardinal veins → common cardinal veins → sinus venosus.

2. Observations on living chick embryos.

Obtain fertile eggs incubated at 38°C for approximately 50-56 hours. These eggs should contain typical 48-hour embryos. Carefully crack open the eggs against the side of a finger bowl half filled with warm saline solution (0.9% NaCl in distilled water). The crack in the shell should be held beneath the surface of the saline solution and the two halves of the shell slowly pulled apart, releasing the egg contents into the solution. Observe particularly the circulation of blood, and especially the passage of blood through the heart. If you watch the living heart you can discern the relationships of its parts much better than in whole mounts. Note the jerkiness of the circulation in the arteries and the smoother flow of blood in the veins. Observe as many as possible of the structures you previously identified in whole mounts. To make identification of most structures easier, apply a drop of a 1% aqueous solution of vital stain (neutral red or nile blue sulfate).

3. Serial transverse sections. Aids: Fig. J; Plates 9-13; A, Figs. 561-574; B, Figs. 264-269, HU, Figs. 113-117; P, Fig. 76.

Position your slide on the microscope stage so that when viewed through the microscope each section is oriented as in Plate 9. Begin tracing sections in antero-posterior sequence under low magnification. Due to the cranial flexure the first brain sections are probably cut through the MESENCEPHALON. These first sections will actually be frontal (that is, cut lengthwise in the right-left plane) rather than transverse. Note the oval shape of the mesencephalon sections as you continue caudally. Sections then begin to lengthen out and a constriction appears in the brain. Sections at this level not only cut across the mesencephalon but also the METENCEPHALON and MYELENCEPHALON (Plate 9, Fig. 1). The constricted region between metencephalon and mesencephalon is the ISTHMUS. The myelencephalon can be readily identified by its thin ROOF PLATE. As you continue to trace sections posteriorly, the brain separates into two parts. The upper part is the myelencephalon (or possibly the metencephalon and myelencephalon; the boundary between these two structures is indistinct); the lower part is now the DIENCEPHALON. At about this level a dark accumulation of cells appears on each side of the body, the SEMILUNAR GANGLIA of the TRIGEMINAL (V) CRANIAL NERVES. These ganglia originate from ectodermal NEURAL CREST CELLS, as well as from a thickening of the lateral skin ectoderm on each side. These ectodermal thickenings are called EPIBRANCHIAL PLACODES; they usually cannot be identified.

Shortly after the brain separates into two parts the NOTOCHORD appears. It is usually cut frontally, due to the cranial flexure, and it quickly separates into two parts, one beneath each brain section. As you continue to follow sections posteriorly, the portion of the notochord beneath the diencephalon disappears. At about this level note an accumulation of cells on each side of the myelencephalon; these are the ACOUSTICO-FACIALIS GANGLIA of the FACIAL (VII) and AUDITORY (VIII) CRANIAL NERVES. The acoustic part of these ganglia originates from the ectodermal AUDITORY PLACODES present earlier. The facialis part originates from both NEURAL CREST CELLS and from a second pair of EPIBRANCHIAL PLACODES (see Plate 10, Fig. 1). These epibranchial placodes may not be readily visible in your embryo. Note the close relationship between the AUDITORY VESICLES and the acoustico-facialis ganglia (Plate 9, Fig. 4; Plate 10, Fig. 1).

Return to the level where the portion of the notochord beneath the diencephalon disappears and trace sections posteriorly. Observe that the diencephalon becomes more and more elongate; the narrow end of the diencephalon which projects towards the myelencephalon is the INFUNDIBULUM. In more posterior sections the OPTIC CUPS appear. In the first sections encountered they are not connected to the diencephalon. The LENS VESICLES appear in more posterior sections and are nestled within the double-layered optic cups (Plate 10, Figs. 1,

2). The layer of each optic cup which is next to the lens vesicle is thickened as the SENSORY RETINA; the much thinner layer is the PIGMENTED RETINA. The space between these two layers, the OPTICOEL, is directly continuous with the cavity of the diencephalon. At about this level a small accumulation of cells appears above each auditory vesicle; these are the SUPERIOR GANGLIA of the GLOSSOPHARYNGEAL (IX) CRANIAL NERVES (Plate 10, Figs. 1, 2). These ganglia originate from NEURAL CREST CELLS.

As you continue to trace sections posteriorly the optic cups become continuous with the diencephalon via the OPTIC STALKS (Plate 10, Fig. 4). In sections posterior to the optic stalks the TELENCEPHALON appears. Note that the skin ectoderm lateral to the telencephalon is thickened as the NASAL (OLFACTORY) PLACODES (Plate 11, Fig. 1), the rudiments of the NASAL CAVITIES. They are induced to form by the adjacent regions of the telencephalon. A few sections more posteriorly the telencephalon disappears.

Return to the level of the auditory vesicles and trace sections posteriorly. The transition from myelencephalon to SPINAL CORD is so gradual that the boundary between the two is indistinguishable. Find a section closely resembling Plate 12, Fig. 4, and examine the spinal cord carefully. Identify the ROOF and FLOOR PLATES. Try to identify NEURAL CREST CELLS. Continue to trace sections posteriorly. As you approach the caudal end of the neural tube it often contains multiple cavities. In sections slightly posterior to this level the notochord appears to fuse to the floor of the neural tube, and a solid mass of cells then appears in the midline. This mass is the TAIL BUD. The caudal portion of the neural tube is formed by cavitation (hollowing out) of a portion of the tail bud, resulting in formation of multiple cavities which subsequently fuse forming a single cavity.

Return to the level where the brain separates into two parts and trace sections posteriorly, noting the appearance of the FOREGUT. It rapidly becomes somewhat triangular in shape (Plate 9, Fig. 4). At this level sections cut through the PHARYNX and FIRST PHARYNGEAL POUCHES. The first pharyngeal pouches project laterally from the pharynx to contact the adjacent regions of the skin ectoderm which they induce to invaginate, forming the FIRST BRANCHIAL GROOVES. The double-layered membranes formed by the endoderm of the pouches and the ectoderm of the grooves are the FIRST CLOSING PLATES.

Return to the first section through the foregut and again trace sections posteriorly. A small vesicle soon appears between the infundibulum and foregut (Plate 9, Fig. 4). This is RATHKE'S POUCH, the rudiment of the ANTERIOR PITUITARY GLAND. Recall that the posterior pituitary gland is formed from the infundibulum. Note the close relationship between the infundibulum and Rathke's pouch. In more posterior sections note that Rathke's pouch opens into a transverse slitlike space, the STOMODEUM (Plate 10, Fig. 1), which is lined by ectoderm. Rathke's pouch develops as an outgrowth from the stomodeum, perhaps due to induction by the infundibulum. The stomodeum lies between two rounded masses next to the foregut, the MANDIBULAR PROCESSES of the first branchial arches, and two rounded or slightly flattened masses on either side of Rathke's pouch, the MAXILLARY PROCESSES of the first branchial arches. Note that the stomodeum dips between the two mandibular processes and is separated from the foregut only by a thin membrane of ectoderm and endoderm, the ORAL MEMBRANE (Plate 10, Fig. 1). This membrane will *later* rupture forming the MOUTH OPENING. The outer part of the mouth is therefore derived from ectoderm of the stomodeum, and the inner part from foregut endoderm. Examine available models to fully understand and visualize these relationships.

Continue tracing sections posteriorly. The pharynx becomes somewhat rounded and then suddenly expands laterally at the level of the SECOND PHARYNGEAL POUCHES (Plate 10, Fig. 4). Identify the SECOND BRANCHIAL GROOVES, if formed, and the SECOND CLOSING PLATES. The considerable depression in the floor of the pharynx at this level is the THYROID RUDIMENT. Continue tracing sections posteriorly and identify the very broad THIRD PHARYNGEAL POUCHES, the THIRD CLOSING PLATES, and the THIRD BRANCHIAL GROOVES, if formed (Plate 11, Fig. 1).

Sections posterior to the third pharyngeal pouches cut through the portion of the FOREGUT caudal to the pharynx. At this level the foregut endoderm is surrounded by splanchnic mesoderm (Plate 11, Fig. 2). In more posterior sections the ventro-lateral portions of the foregut may be slightly evaginated. These evaginations are the beginnings of the LUNG BUDS (Plate 11, Fig. 3). The paired portions of the coelom lateral to the developing lung buds and continuous with the PERICARDIAL CAVITY are the PLEURAL CAVITIES. Caudal to this level the pleural and pericardial cavities are separate (Plate 11, Fig. 4).

At about this level notice a small mass of cells ventral to the foregut; this is the CRANIAL LIVER RUDIMENT. Continue caudally noting that the cranial liver rudiment becomes continuous with the foregut (Plate 12, Fig. 1). This region of the foregut is the DUODENUM. Your embryo may also contain a CAUDAL LIVER RUDIMENT. If so, it will appear as a branch (or as branches) of the ventral part of the cranial liver rudiment. The paired portions of the coelom lateral to the duodenum are the RUDIMENTS OF THE PERITONEAL CAVITY (Plate 12, Fig. 1). These paired rudiments continue caudally for many sections and *later* fuse, forming a single PERITONEAL CAVITY surrounding the gut. As sections are traced posteriorly from the level of the duodenum, the CRANIAL INTESTINAL PORTAL is encountered almost immediately. More caudally the embryo progressively flattens dorso-ventrally, and endoderm forms its lower layer (Plate 12, Fig. 4). In more posterior sections an endoderm-lined cavity, the ALLANTOIS RUDIMENT, may appear beneath the tail bud (Plate 13, Fig. 2). The exact mechanism by which this rudiment forms is unknown. It will *later* greatly enlarge and become covered externally with splanchnic mesoderm forming the ALLANTOIS, one of the EXTRA-EMBRYONIC MEMBRANES. In slightly more *anterior* sections the floor of the allantois rudiment disappears; this is the region of the CAUDAL INTESTINAL PORTAL (Plate 13, Fig. 1). Thus the allantois rudiment opens into the SUBGERMINAL CAVITY.

Return to the level of the first somite pair and trace sections posteriorly. The structure of the cranial SOMITES is now quite different from that of the 33-hour stage. Examine them closely (Plate 11, Figs. 2, 3). Each somite is now subdivided into three parts: DERMATOME (a plate of darkly stained cells lying just beneath the skin ectoderm – source of some of the DERMIS); MYOTOME (a plate of lightly stained cells lying just medial to the dermatome – source of some of the SKELETAL MUSCLES); SCLEROTOME (a diffuse mass of mesenchymal cells lying between myotome, neural tube, notochord, and dorsal aorta or descending aorta – source of the VERTEBRAL COLUMN and RIBS). The manner of formation of these somite subdivisions may be studied by comparing the most caudal somites, in which the subdivision is just beginning, with more cranial somites.

Caudal to the last somite pair identify the SEGMENTAL PLATES and note that their caudal ends are continuous with the TAIL BUD. The tail bud is a mass of mesenchymal cells located at the caudal end of the embryo. It contributes cells which form the neural tube, segmental plates, and mesenchyme of the tail.

Return to the level of the cranial liver rudiment (Plate 11, Fig. 4) and trace sections posteriorly. Lateral to the descending aorta or dorsal aortae identify the paired MESONEPHRIC DUCT RUDIMENTS (Plate 12, Figs. 2-4). They were initially solid longitudinal structures (see page 10), but now are undergoing cavitation in most embryos. Medial to the mesonephric duct rudiments identify the usually solid MESONEPHRIC TUBULE RUDIMENTS. The latter are induced to form from the NEPHROTOME by the mesonephric duct rudiments. The mesonephric tubule rudiments undergo cavitation to form the MESONEPHRIC TUBULES of the paired MESONEPHRIC KIDNEYS.

Return to the first section through the foregut (Plate 9, Fig. 3). On each side of the foregut is a large blood vessel cut frontally, the DORSAL AORTA. Trace sections posteriorly and note that two vessels appear in the place of each dorsal aorta (Plate 9, Fig. 4). The one which lies above the first pharyngeal pouch is still the dorsal aorta; the one which lies beneath the first pouch is the FIRST AORTIC ARCH (see Plate 8, Fig. 2). As sections are traced further posteriorly, note that the first aortic arches become located within the mandibular processes (Plate 10, Figs. 1, 2). Continue to trace sections posteriorly until the first aortic arches approach one another and become continuous with the AORTIC SAC. Continue tracing the aortic sac caudally. Just caudal to the first pharyngeal pouches note a vessel extending downwards from each dorsal aorta. These vessels are the SECOND AORTIC ARCHES (Plate 10, Figs. 2-4). A few sections more posteriorly their ventral ends become continuous with the aortic sac. The second aortic arches are contained within the second branchial arches. Continue tracing sections posteriorly. Just caudal to the second pharyngeal pouches another pair of downward extensions from the dorsal aortae may be seen; these are the THIRD AORTIC ARCHES which are usually just developing. More caudally the ventral ends of these vessels become continuous with the aortic sac. The third aortic arches are contained within the third branchial arches.

Follow the aortic sac caudally noting the appearance of ENDOCARDIUM clearly separated from EPI-MYOCARDIUM, indicating the beginning of the BULBUS CORDIS of the heart (Plate 11, Fig. 1). In more posterior sections identify the VENTRICLE and ATRIUM (Plate 11, Fig. 2). Recall that these regions lie *beneath* the bulbus cordis in surface view. Notice that the endocardium of the atrium is in close contact with the epi-myocardium, whereas these two layers are widely separated in the ventricle and bulbus cordis. At about this level note that the two dorsal aortae are fused as the DESCENDING AORTA. In more posterior sections a portion of the heart is attached to the foregut by the DORSAL MESOCARDIUM (Plate 11, Fig. 3). This portion is the SINUS VENOSUS. At about this level the bulbus cordis and ventricle become continuous. Slightly more caudally the sinus venosus is continuous with a blood vessel on each side; these vessels are the COMMON CARDINAL VEINS (Plate 11, Fig. 4).

Continue caudally and note that the sinus venosus appears to divide into two vessels, one on each side of the cranial liver rudiment (Plate 12, Fig. 1). These two vessels are the VITELLINE VEINS. Trace the vitelline veins caudally, noting that they shift laterally within the splanchnic mesoderm. Shift your attention to the descending aorta and trace it caudally. Identify small blood vessels continuous with the descending aorta dorsally; these are the INTERSEGMENTAL ARTERIES (see Plate 12, Fig. 1). Note that at more caudal levels the descending aorta has not yet formed and paired DORSAL AORTAE are again present. Continue to trace the dorsal aortae caudally. They become continuous with the VITELLINE ARTERIES which extend laterally within the splanchnic mesoderm (Plate 12, Fig. 4).

Return to the section where the common cardinal veins are continuous with the sinus venosus (Plate 11, Fig. 4). Each common cardinal vein is contained within a bridge of mesoderm connecting the lateral body wall to the sinus venosus. Trace sections *anteriorly,* observing that the lateral body walls separate from the heart (Plate 11, Fig. 3). Note a vessel located within the lateral body wall on each side near the heart at this level. In more *anterior* sections these vessels are located progressively more dorsally within the lateral body walls. These vessels are still the COMMON CARDINAL VEINS near the heart (Plate 11, Fig. 3), but they are the PRECARDINAL VEINS at a level approximately lateral to the foregut (Plate 11, Fig. 2). Trace sections *anteriorly.* The precardinal veins lie progressively more dorsally and come to lie ventro-lateral to the myelencephalon (Plate 10, Fig. 2). In more *anterior* sections they are cut frontally due to the cranial flexure (Plate 9, Fig. 2).

Return to a level similar to Plate 11, Fig. 1 and trace sections *posteriorly, carefully* observing the precardinal veins. They become located progressively more ventrally and suddenly, at about a level lateral to the foregut, each appears to separate into two vessels. The dorso-most vessel, usually very small, is the POSTCARDINAL VEIN; the ventro-most vessel is the COMMON CARDINAL VEIN (Plate 11, Fig. 3). See Fig. J. Continue to trace sections posteriorly and note again that the common cardinal veins become continuous with the sinus venosus. The postcardinal veins become located progressively more dorsally and eventually lie dorsal to the developing kidneys (Plate 12, Figs. 3, 4). In some embryos small blood vessels continuous with the postcardinal veins dorsally can be identified; these are the INTERSEGMENTAL VEINS.

Four EXTRA-EMBRYONIC MEMBRANES are in the process of formation. Two originate from splanchnopleure, the ALLANTOIS and YOLK SAC, and two form somatopleure, the AMNION and CHORION.

The development of the ALLANTOIS is barely initiated. Only a small endoderm-lined cavity, the ALLANTOIS RUDIMENT, has formed (Plate 13, Fig. 2). The allantois later stores nitrogenous waste products and eventually fuses with the chorion to form the CHORIO-ALLANTOIC MEMBRANE (see below).

The YOLK SAC is formed from the splanchnopleure which overgrows the yolk. It lies between the *left* side of the embryo and the yolk at cranial levels where TORSION is complete (Plate 9, Fig. 2; Plate 10, Fig. 4). At these levels the yolk sac is separated from the embryo due to the action of the splanchnopleural component of the LATERAL BODY FOLDS. More caudally the yolk sac has not yet separated from the developing gut (Plate 12, Fig. 2). Further caudally TORSION has not yet occurred and the ventral surface of the embryo as well as the yolk sac lie above the yolk (Plate 12, Fig. 3). Notice the numerous VITELLINE BLOOD VESSELS within the splanchnic mesoderm of the yolk sac. The endodermal cells of the yolk sac digest the yolk, the products of this digestion enter the plexus of vitelline blood vessels, and these products are transported to the developing embryo via the vitelline veins.

The AMNION and CHORION are formed simultaneously by elevation and fusion of amniotic folds. Select a section resembling Plate 12, Fig. 3, showing the LATERAL AMNIOTIC FOLDS. Note that the amniotic folds consist of somatopleure and are continuous with the lateral body walls. The outer wall of each fold consists of *ectoderm on the outside* with an adjacent layer of *somatic mesoderm on the inside,* and forms the CHORION. The inner wall of each fold consists of *somatic mesoderm on the outside* and *ectoderm on the inside,* and forms the AMNION. The somatic mesoderm-lined cavity within each fold is the EXTRA-EMBRYONIC COELOM; it is continuous with the INTRA-EMBRYONIC COELOM (that is, the pericardial, peritoneal, and pleural cavities). The extra- and intra-embryonic coeloms become separated due to the action of the somatopleural component of the LATERAL BODY FOLDS. The cavity developing between the amnion and embryo is the ectoderm-lined AMNIOTIC CAVITY. Trace sections *anteriorly* and observe the region of fusion of the amniotic folds. In more anterior sections the outer chorion is separated from the inner amnion. The amniotic cavity is later filled with AMNIOTIC FLUID. The embryo then floats freely within this protective fluid. The extra-embryonic coelom later contains the enlarging allantois. Wherever the splanchnic mesoderm of the allantois contacts the somatic mesoderm of the chorion fusion occurs forming the CHORIO-ALLANTOIC MEMBRANE. The latter lies immediately beneath the inner shell membrane and functions in respiration and absorption of calcium from the shell.

4. Serial sagittal sections. Aids: Plate 13, Figs. 3, 4; M, Fig. 108.

Only about the cranial one-third of the embryo is actually cut sagittally; the remaining caudal regions are cut frontally. Do not waste time looking for particular structures. Instead, try to find a section which is approximately a midsagittal section through the BRAIN and NOTOCHORD. Identify all regions of the brain including the INFUNDIBULUM. Also identify the PHARYNX, ORAL MEMBRANE, and PREORAL GUT (SEESSEL'S POUCH). The latter is that part of the foregut which lies cranial to the oral membrane; it later degenerates. Identify the ectodermal STOMODEUM and its evagination, RATHKE'S POUCH, which extends towards the infundibulum.

Focus your attention on parasagittal sections in which you can recognize the following structures (Plate 13, Fig. 4): OPTIC CUP (including the SENSORY RETINA and PIGMENTED RETINA), LENS VESICLE, and OPTIC FISSURE. Identify the OPTIC STALK connecting the optic cup to the DIENCEPHALON. The optic fissure is formed by ventral invagination of the optic cups and stalks. At this stage the optic fissure is probably formed only in the optic cup, or at most in the optic cup and in the part of the optic stalk immediately adjacent to the optic cup. The optic fissure is important in development because OPTIC NERVE FIBERS from the retina grow back to their proper connections within the diencephalon through the wall of the fissure. Moreover, the cavity of the fissure provides a pathway for blood vessels to enter the optic cup. Later this fissure will close.

5. Summary of germ layer derivatives formed since the 33-hour stage.

Ectoderm	*Mesoderm*	*Ectoderm and Somatic Mesoderm*
auditory vesicles	dermatomes	amnion
branchial grooves	mesonephric duct rudiments	chorion
cranial ganglia:	mesonephric tubule rudiments	
V. semilunar	myotomes	
VII.-VIII. acoustico-facialis	sclerotomes	*Endoderm and Splanchnic Mesoderm*
IX. superior		
epibranchial placodes		yolk sac
lens vesicles	*Endoderm*	
nasal placodes		
optic cups	allantois rudiment	*Ectoderm and Endoderm*
optic stalks	cranial liver rudiment	
Rathke's pouch	caudal liver rudiment	closing plates
stomodeum	duodenum	
	lung buds	
	pharynx	*Ectoderm, Mesoderm, and Endoderm*
	pharyngeal pouches	
	preoral gut	branchial arches

K. 72-HOUR CHICK EMBRYOS

1. **Whole mounts: injected and uninjected. Aids:** Plates 14-16; A, Figs. 575, 577; HU, Fig. 130; M, Figs. 122-126.

First examine your uninjected whole mount slide with the naked eye. Superimpose this slide on the 48-hour uninjected whole mount slide and note the changes that have occurred in the shape of the embryo. In addition to the CRANIAL FLEXURE two other flexures are probably well developed. These are the CERVICAL FLEXURE, at the level of the first somites (Plate 14, Fig. 1), and the TAIL FLEXURE, at the caudal end of the embryo (Plate 14, Fig. 2). Note that the embryo is sharply bounded and that it lies on its *left* side throughout a considerable part of its length. This means that it has undergone (or is undergoing) TORSION throughout most of its length. Somewhat encircled by the TAIL is a saclike structure, the ALLANTOIS (Plate 14, Figs. 2, 3). Paired WING and LEG BUDS can be identified, but they are not sharply bounded (Plate 14, Figs. 1, 2). Determine the approximate somite levels at which the limb buds lie. The entire embryo is usually completely covered by AMNION and CHORION, although the oval or circular boundary of the continuous CRANIAL, LATERAL, and CAUDAL (POSTERIOR) AMNIOTIC FOLDS may be visible (Plate 14, Fig. 1).

Examine your slide under low power and identify the following structures: TELENCEPHALON (with its lateral oval-shaped expansions, the CEREBRAL HEMISPHERES); NASAL PITS (formed by invagination of the nasal placodes); DIENCEPHALON; PINEAL GLAND or EPIPHYSIS (a small dorsal evagination of the diencephalon); OPTIC CUPS; OPTIC FISSURES; LENS VESICLES; INFUNDIBULUM; RATHKE'S POUCH (examine under high power and note its close relationship to the infundibulum; see Plate 15); MESENCEPHALON; METENCEPHALON; ISTHMUS; MYELENCEPHALON (note its thin ROOF PLATE); AUDITORY VESICLES; ENDOLYMPHATIC DUCTS (formed by evagination of the dorsal part of the auditory vesicles); SPINAL CORD; CRANIAL INTESTINAL PORTAL; and CAUDAL INTESTINAL PORTAL, if visible.

Just cranial to the auditory vesicles identify the ACOUSTICO-FACIALIS GANGLIA (Plate 14, Fig. 2). They lie above the SECOND BRANCHIAL ARCHES. The facialis part of these ganglia is the source of the SENSORY NERVE FIBERS of the FACIAL NERVES which innervate the *second* branchial arches. Identify the SEMILUNAR GANGLIA which lie above the FIRST BRANCHIAL ARCHES (Plate 14, Fig. 2). These ganglia are the source of SENSORY NERVE FIBERS of the TRIGEMINAL NERVES which innervate the *first* branchial arches.

Observe that the first branchial arches are partially split by the STOMODEUM into MAXILLARY PROCESSES (which will form the lateral portions of the UPPER JAW and most of the CHEEKS), and MANDIBULAR PROCESSES (which will form the LOWER JAW). The maxillary processes may partially overlap the mandibular processes in some embryos. Identify the FIRST BRANCHIAL GROOVES, just caudal to the mandibular processes, the SECOND BRANCHIAL GROOVES, between the SECOND and THIRD BRANCHIAL ARCHES, and the THIRD BRANCHIAL GROOVES.

In favorably injected embryos the PRECARDINAL, POSTCARDINAL, and COMMON CARDINAL VEINS, as well as the INTERSEGMENTAL VEINS (continuous with the post-cardinal veins dorsally) can be identified. The common cardinal veins return blood to the SINUS VENOSUS. Identify the following structures (see Plate 14, Figs. 1-3; Plate 16, Figs. 1, 2): ATRIUM; VENTRICLE; BULBUS CORDIS; AORTIC SAC; FIRST, SECOND, THIRD, and possibly FOURTH AORTIC ARCHES (the first aortic arches are undergoing degeneration and may be reduced in diameter); DORSAL AORTAE; INTERNAL CAROTID ARTERIES (cranial extensions of the dorsal aortae); DESCENDING AORTA; and INTERSEGMENTAL ARTERIES. Also identify the VITELLINE ARTERIES, which are continuous with the dorsal aortae at approximately the 22nd somite level, and VITELLINE VEINS. The latter are fused near the level of the heart to form a single vessel, the DUCTUS VENOSUS, which is continuous with the sinus venosus. Try to identify the ductus venosus.

2. Observations on living chick embryos.

Obtain fertile eggs incubated at 38°C for approximately 72 hours. These eggs should contain typical 72-hour embryos. Open and examine these eggs exactly as described on page 27.

3. Serial transverse sections. Aids: Plates 16-22; A, Figs. 578-588; HU, Figs. 131-136b; M, Figs. 130-142.

Position your slide on the microscope stage so that when viewed through the microscope each section is oriented as in Plate 16, Fig. 3. Trace sections in antero-posterior sequence unless otherwise specified. Due to the CERVICAL FLEXURE the first brain sections are probably cut through the MYELENCEPHALON (and possibly through the METENCEPHALON as well). A variable amount of tissue may appear to be lying freely within the cavity of the myelencephalon. This tissue is the thin ROOF PLATE of the myelencephalon and the adjacent SKIN ECTODERM. Continue tracing sections posteriorly. ENDOLYMPHATIC DUCTS soon appear alongside the walls of the myelencephalon, and a few sections more posteriorly these are continuous with the AUDITORY VESICLES. At about this level, sections begin to cut across a small, rounded region lying beneath the metencephalon. This region is the MESENCEPHALON; it is continuous with the metencephalon in more posterior sections (Plate 16, Fig. 3). Note the segmental enlargements of the walls of the rhombencephalon; these are the NEUROMERES.

As sections are traced posteriorly from about the level at which the auditory vesicles are first identified, two groups of nerve fibers appear to emerge from neuromeres cranial to each auditory vesicle. They quickly become continuous with two prominent ganglia on each side (Plate 16, Fig. 3). The ganglia lying against the cranial wall of the auditory vesicles are the ACOUSTICO-FACIALIS GANGLIA (future ACOUSTIC and GENICULATE GANGLIA). The other very large ganglia are the SEMILUNAR GANGLIA. At about this level note on the caudal side of each auditory vesicle a poorly circumscribed, very small, rounded accumulation of cells, the SUPERIOR GANGLIA. In slightly more posterior sections the neural tube separates into two parts. The upper part is either the myelencephalon or the SPINAL CORD, while the lower part is primarily the mesencephalon.

As you continue tracing sections posteriorly the superior ganglia are directly continuous, without a distinct boundary, with SENSORY NERVE FIBERS of the GLOSSOPHARYNGEAL

(IX) NERVES (Plate 17, Fig. 1). Similarly, the acoustico-facialis ganglia are directly continuous with the SENSORY NERVE FIBERS of the FACIAL (VII) NERVES. The facial nerves can be traced into the SECOND BRANCHIAL ARCHES, where they merge on each side with a distinct EPIBRANCHIAL PLACODE. These placodes contribute cells to the acoustico-facialis ganglia. Also trace the glossopharyngeal nerves into the THIRD BRANCHIAL ARCHES where they likewise merge on each side with an EPIBRANCHIAL PLACODE (Plate 17, Fig. 3). These placodes are the sources of the PETROSAL GANGLIA of the glossopharyngeal nerves. *If you trace the facial and glossopharyngeal nerves into their respective branchial arches, correct identification of these arches becomes relatively simple.* At about the level where each glossopharyngeal nerve merges with an epibranchial placode try to identify an accumulation of NEURAL CREST CELLS on each side, the JUGULAR GANGLION of the VAGUS (X) CRANIAL NERVE (Plate 17, Fig. 3). The jugular ganglia fade out within the FOURTH BRANCHIAL ARCHES.

Return to sections through the semilunar ganglia and note that each one appears to subdivide caudally into three branches, the branches of the TRIGEMINAL (V) NERVE (Plate 17, Fig. 1). The medial branch is the MAXILLARY BRANCH; the lateral branch closest to the mesencephalon is the OPHTHALMIC BRANCH; the lateral branch above the ophthalmic is the MANDIBULAR BRANCH. These branches are contained within the region of the FIRST BRANCHIAL ARCHES and fade out caudally.

Shift your attention to the mesencephalon and follow it posteriorly in your sections (cranially within the embryo). Nerve fibers (AXONS) appear to emerge from the floor of the mesencephalon to either side of the midline. These will probably *not* be cut symmetrically. The accumulations of NEURAL ECTODERMAL CELLS from which these nerve fibers arise lie within the floor of the mesencephalon. Each accumulation is designated as a MOTOR NUCLEUS. MOTOR NERVE FIBERS grow out from the motor nuclei towards areas they will later innervate; collectively these nerve fibers (axons) constitute the OCULOMOTOR (III) CRANIAL NERVES. These nerves will *later* innervate four pairs of extrinsic EYE MUSCLES.

In establishment of *sensory* portions of cranial nerves the following events occur: (1) Cranial ganglia, consisting of CELL BODIES of YOUNG SENSORY NEURONS, are formed adjacent to the brain from accumulations of ectodermal neural crest cells and/or cells derived from ectodermal placodes. (2) Sensory nerve fibers, AXONS, grow out from these ganglia and into the brain wall. (3) Other sensory nerve fibers, DENDRITES, grow out from these ganglia towards the areas to be innervated.

In establishment of *motor* portions of cranial nerves the following events occur: (1) Motor nuclei, consisting of CELL BODIES of YOUNG MOTOR NEURONS, are formed within the ventral part of the wall of the brain from accumulations of neural ectodermal cells. (2) Motor nerve fibers, AXONS, grow out from these nuclei, leave the wall of the brain, and extend towards the areas to be innervated. (3) Other motor nerve fibers, DENDRITES, grow out from these nuclei but remain within the brain wall.

The chart on the following page summarizes the development of cranial nerves through the 72-hour stage.

In sections posterior to the oculomotor nerves, the INFUNDIBULUM appears, marking the beginning of the DIENCEPHALON. Identify the OPTIC CUPS and the OPTIC FISSURES, the regions where the ventral lip of each optic cup is partially or completely lacking (Plate 19, Fig. 2). The LENS VESICLE is now completely closed. It is separated from the overlying SKIN ECTODERM which *later* forms the CORNEAL EPITHELIUM (Plate 19, Fig. 1). The lens vesicle is beginning to differentiate into two regions: the outer thin LENS EPITHELIUM; the inner thick LENS FIBERS. Continue to trace sections posteriorly until sections cut through both the diencephalon and TELENCEPHALON (Plate 20, Fig. 2). The telencephalon is expanded laterally as the CEREBRAL HEMISPHERES (Plate 20, Figs. 2, 3). At about this level identify the NASAL PITS and PINEAL GLAND (Plate 20, Fig. 3). Continue tracing sections posteriorly noting the disappearance of the diencephalon and telencephalon.

Quickly follow the SPINAL CORD throughout its length. NEURAL CREST CELLS are now aggregated alongside its most cranial levels forming the paired SPINAL GANGLIA.

Cranial Nerves	*Cranial Ganglia Present*	*Origin of Cranial Ganglia or Motor Nuclei*	*Type of Nerve Fibers Present*	*Regions Innervated*
oculomotor	———	neural ectoderm	motor	four pairs of extrinsic eye muscles
trigeminal	semilunar	neural crest cells and epibranchial placodes	sensory (*later* also motor)	first branchial arches
facial	acoustico-facialis (*later* geniculate)	neural crest cells and epibranchial placodes	sensory (*later* also motor)	second branchial arches
auditory (nerves not yet formed)	acoustico-facialis (*later* acoustic)	auditory placodes	(*later* sensory)	inner ears
glossopharyngeal	superior	neural crest cells	sensory (*later* also motor)	third branchial arches
vagus (nerves not yet formed)	jugular	neural crest cells	(*later* sensory and motor)	fourth branchial arches

Segmentation of these ganglia corresponds to segmentation of the SOMITES, which are the much more prominent and elongated structures lateral to each ganglion. Spinal ganglia are located beneath the MYOTOME in the middle of the cranio-caudal extent of each somite. SPINAL NERVES will *later* form. Examine the most caudal tip of the spinal cord, observing that the NOTOCHORD is still distinguishable at this level. The spinal cord and notochord are usually cut frontally at this level due to the TAIL FLEXURE.

Return to the most anterior section in which the notochord can be identified (Plate 17, Fig. 1). At this level it is cut frontally due to the CERVICAL FLEXURE. Trace sections posteriorly. Sections soon cut through the FIRST, and possibly SECOND, PHARYNGEAL POUCHES, which at first appear as isolated structures (Plate 17, Fig. 2). As you continue to trace sections posteriorly, both pairs of pouches become continuous with the PHARYNX (Plate 17, Fig. 3). (Remember to use the facial nerves as a landmark for identification of the second branchial arches. Using the facial nerves as a landmark will enable you to identify the first and second pharyngeal pouches with certainty.) Both the first and second pharyngeal pouches may open to the outside via the FIRST and SECOND BRANCHIAL CLEFTS, which arise by rupture of the corresponding CLOSING PLATES. Identify the shallow FIRST and SECOND BRANCHIAL GROOVES. In more posterior sections the THIRD PHARYNGEAL POUCHES appear and almost immediately become continuous with the pharynx (Plate 18, Fig. 1). Identify the THIRD CLOSING PLATES and the shallow THIRD BRANCHIAL GROOVES. Note that the first pharyngeal pouches fade out at about this level. In more posterior sections identify the FOURTH PHARYNGEAL POUCHES, FOURTH CLOSING PLATES, FOURTH BRANCHIAL GROOVES (very shallow), and THYROID RUDIMENT (Plate 18, Figs. 2, 3).

Return to a section similar to Plate 18, Fig. 1 and identify the PREORAL GUT, which is located just cranial to the pharynx. Also identify the FIRST BRANCHIAL ARCHES, just lateral to the preoral gut. In *slightly* more posterior sections the STOMODEUM appears, and separates the first branchial arches into MAXILLARY and MANDIBULAR PROCESSES. At about this level the cavity of the gut usually opens into the stomodeum via the MOUTH OPENING, since the ORAL MEMBRANE is probably ruptured. Try to identify the mouth opening in your embryo. It will be seen in sections located between those illustrated by Plate 18, Figs. 1 and 2. In slightly more posterior sections identify RATHKE'S POUCH

(Plate 18, Fig. 2), which quickly becomes continuous with the stomodeum (Plate 18, Fig. 3). Note the close relationship between Rathke's pouch and INFUNDIBULUM.

Continue tracing sections posteriorly and identify the LARYNGO-TRACHEAL GROOVE, which is continuous, without a distinct boundary, with the pharynx (Plate 19, Fig. 1). (The laryngo-tracheal groove is formed by elongation of the ventral portion of the foregut, at the level where lung bud formation was initiated earlier. See Plate 11, Fig. 3.) More posterior sections cut through the LUNG BUDS, which at this stage are bilateral expansions of the laryngo-tracheal groove (Plate 19, Fig. 3). Moreover, the laryngo-tracheal groove is beginning to constrict between the pharynx and lung buds. In slightly more posterior sections this constriction is complete, and paired lung buds lie beneath the ESOPHAGUS (Plate 20, Fig. 1). Both the esophagus and lung buds are contained within a thick mesentery composed of splanchnic mesoderm. The portion of the mesentery dorsal to the esophagus is the MESOESOPHAGUS; the portion dorsal to the sinus venosus is the DORSAL MESOCARDIUM. The coelomic cavities lateral to the lung buds are the PLEURAL CAVITIES.

Continue tracing sections posteriorly and note the disappearance of the lung buds. At about this level two narrow coelomic cavities are located ventro-lateral to the esophagus. These cavities are extensions of the pleural cavities (Plate 20, Fig. 2). In slightly more posterior sections they become continuous with the more lateral portions of the pleural cavities (Plate 20, Fig. 3). At about this level the diameter of the gut is usually slightly larger; this is the region of the STOMACH. The mesentery dorsal to the stomach is the DORSAL MESOGASTER; the one ventral to it is the GASTRO-HEPATIC LIGAMENT (VENTRAL MESOGASTER). Although the stomach lies medial to the pleural cavities at this stage, it will *later* descend and be enclosed by the PERITONEAL CAVITY.

In sections at about the level of the stomach identify the CRANIAL LIVER RUDIMENT (Plate 20, Fig. 3); it lies just above a large blood vessel, the DUCTUS VENOSUS. In more posterior sections the cranial liver rudiment and gut become continuous (Plate 21, Fig. 1). This level of the gut is the DUODENUM. The mesentery lying dorsal to the duodenum is the MESODUODENUM; the one ventral to it is the DUODENO-HEPATIC LIGAMENT. At about this level identify the CAUDAL LIVER RUDIMENT lying beneath the ductus venosus. The close relationship between the ductus venosus and liver tissue is very important. *Later,* as the liver tissue grows, it invades the ductus venosus and subdivides this vessel into smaller channels, the HEPATIC SINUSOIDS.

Continue tracing sections posteriorly and notice that the cranial liver rudiment lengthens dorso-ventrally and becomes continuous with the caudal liver rudiment (Plate 21, Fig. 2). The cranial liver rudiment lies between two blood vessels at this level, the VITELLINE VEINS. The boundaries between the duodenum, cranial liver rudiment, and caudal liver rudiment are indistinct. At this level note a solid mass of cells continuous with the dorsal wall of the duodenum; this is the DORSAL PANCREATIC RUDIMENT. In more posterior sections identify the CRANIAL INTESTINAL PORTAL (Plate 21, Fig. 3).

Trace sections posteriorly and identify the paired WING BUDS (Plate 22, Fig. 1), and further posteriorly, the paired LEG BUDS (Plate 22, Figs. 2, 3). Both wing and leg buds consist of a core of somatic mesoderm covered by skin ectoderm, which is thickened laterally as the APICAL ECTODERMAL RIDGE. Experiments suggest that this ridge is necessary for normal outgrowth and development of limb buds.

In sections through about the level of the leg buds, identify an endoderm-lined cavity, the HINDGUT (Plate 22, Fig. 2), which is formed by the combined action of the TAIL FOLD OF THE BODY and LATERAL BODY FOLDS. In more posterior sections a portion of the hindgut, the CLOACA, is continuous ventrally with the ALLANTOIS (Plate 22, Fig. 3). The allantois is formed by expansion of the endodermal allantois rudiment which has become surrounded by splanchnic mesoderm. Trace the cloaca and allantois caudally noting the disappearance of the latter, and the separation of the TAIL from AMNION by the ectoderm-lined SUBCAUDAL POCKET (Plate 22, Fig. 4). At about this level the ventral endoderm of the cloaca usually contacts the ventral skin ectoderm forming the double-layered CLOACAL MEMBRANE.

Caudal to this membrane sections usually cut through the much smaller TAIL GUT, which will soon degenerate.

Return to a section similar to Plate 22, Fig. 2 and trace sections *anteriorly,* noting the disappearance of the floor of the HINDGUT. This is the level of the CAUDAL INTESTINAL PORTAL, the opening of the hindgut into the ~~SUBGERMINAL CAVITY~~. (Due to the action of the TAIL FOLD OF THE BODY the allantois rudiment has been displaced ventrad. Therefore the allantois opens into the floor of the cloaca rather than into the subgerminal cavity.)

Return cranially to a level showing well-formed SOMITES and identify DERMATOME, MYOTOME, and SCLEROTOME (Plate 22, Fig. 1). Trace sections *posteriorly* noting that tail somites are considerably less advanced in their differentiation. Caudal to the somites identify the short SEGMENTAL PLATES, whose caudal ends are continuous with the small TAIL BUD.

Return to the level where the cranial liver rudiment connects to the duodenum (Plate 21, Fig. 1). At about this level a tiny duct can be seen on each side, dorso-lateral to the meso-duodenum. These ducts are the MESONEPHRIC DUCTS, which formed by cavitation of the mesonephric duct rudiments. The cranial end of each mesonephric duct is undergoing degeneration. Trace sections posteriorly and observe that MESONEPHRIC TUBULES are forming from MESONEPHRIC TUBULE RUDIMENTS, medial to each mesonephric duct (Plate 22, Fig. 1). Some mesonephric tubules may have established connections with the mesonephric ducts. The mesonephric tubules collectively form the paired MESONEPHRIC KIDNEYS. Continue tracing sections posteriorly. Observe that the mesonephric ducts have extended caudad beyond the developing mesonephric tubules (Plate 22, Figs. 2, 3), and that they either join the cloaca laterally or terminate near it. Slightly later than the 72-hour stage the mesonephric tubules become functional and extract nitrogenous wastes from the blood. These wastes pass through the mesonephric tubules and ducts and then to the cloaca; they then enter the allantois where they are stored.

Return to the section in which the notochord is cut frontally (Plate 17, Fig. 1) and trace sections posteriorly. Identify the paired DORSAL AORTAE (Plate 17, Fig. 2). Due to the CERVICAL FLEXURE they are at first cut frontally. Each dorsal aorta appears to be constricted into two parts at the level of the first pharyngeal pouches. The lower part is the INTERNAL CAROTID ARTERY, the cranial extension of the dorsal aorta. Quickly trace sections posteriorly and observe that the internal carotid arteries fade out at about the level of the diencephalon.

Return to a level similar to Plate 17, Fig. 2 and trace sections posteriorly. Observe that the SECOND AORTIC ARCHES extend from the dorsal aortae into the second branchial arches, and subsequently the THIRD AORTIC ARCHES extend into the third branchial arches (Plate 17, Fig. 3). If the FIRST AORTIC ARCHES are still present in your embryo, note that they extend from the internal carotid arteries into the first branchial arches. In more posterior sections the small FOURTH AORTIC ARCHES extend from the dorsal aortae into the fourth branchial arches (Plate 18, Fig. 2). At about this level note that the dorsal aortae are fused as the DESCENDING AORTA.

Continue tracing sections posteriorly observing the aortic arches. In the region of the thyroid gland they approach one another (Plate 18, Fig. 3) and soon unite with the AORTIC SAC (see Plate 19, Fig. 1). In more posterior sections identify the BULBUS CORDIS, ATRIUM, which lies to the *left* (apparent right) of the bulbus cordis, and SINUS VENOSUS, which lies in the midline (Plate 19, Fig. 3). In more posterior sections the sinus venosus is continuous with the COMMON CARDINAL VEINS. Usually the *right* (apparent left) common cardinal vein is continuous with the sinus venosus more cranially than the *left* common cardinal vein (see Plate 20, Fig. 2). At about this level sections begin to cut across a small portion of the VENTRICLE, which is continuous with the atrium.

Continue to trace sections posteriorly and observe the gradual disappearance of the atrium, and the progressive enlargement of the ventricle in its place (Plate 20, Fig. 3). Also note the gradual transition between sinus venosus and DUCTUS VENOSUS. Further caudally the bulbus cordis and ventricle become continuous, without a distinct boundary, and the ductus venosus appears to separate into two vessels, the VITELLINE VEINS, which extend laterally into the YOLK SAC (Plate 21). Trace sections posteriorly noting the gradual disappearance of the ventricle.

Return to the level where the DESCENDING AORTA first appears (Plate 18, Fig. 1) and trace sections posteriorly. Identify small blood vessels, the INTERSEGMENTAL ARTERIES, continuous with the descending aorta dorsally (Plate 18, Fig. 2). Quickly trace sections posteriorly until paired DORSAL AORTAE again appear (usually between the levels of the wing and leg buds). Just caudal to this level the VITELLINE ARTERIES are continuous with the dorsal aortae. The vitelline arteries extend laterally into the YOLK SAC; at this level they lie beneath (ventral to) the VITELLINE VEINS. Continue tracing the dorsal aortae caudally and note that they become markedly reduced in diameter (Plate 22, Fig. 2). Within the tail region (that is, caudal to the leg buds) they lie just dorsal to the gut and are designated as CAUDAL ARTERIES.

Return to the level where the *right* (apparent left) COMMON CARDINAL VEIN is continuous with the SINUS VENOSUS (Plate 20, Fig. 2). The right common cardinal vein, as at the 48-hour stage, is contained within a bridge of mesoderm connecting the lateral body wall to the sinus venosus. Trace sections *anteriorly,* observing that the lateral body wall separates from the heart. Two vessels are usually located within the right lateral body wall at this level (Plate 20, Fig. 1). The lower vessel, nearest the sinus venosus, is still the common cardinal vein; the upper vessel, lateral to the descending aorta, is the POSTCARDINAL VEIN. In more *anterior* sections these vessels approach one another and become directly continuous (Plate 19, Figs. 2, 1). At this level these vessels are cut *frontally.* (In some embryos the common cardinal and postcardinal veins may be interconnected by smaller vessels at some levels and completely separated in more anterior sections. If this is the case in your embryo, trace sections anteriorly until the common cardinal and postcardinal veins are broadly continuous.) Continue to trace sections *anteriorly,* noting the disappearance of the upper portion of this frontally cut vessel. The lower persisting portion is now the PRECARDINAL VEIN. In more *anterior* sections the precardinal vein first lies ventro-lateral to the descending aorta (Plate 18, Fig. 3), then lateral to the dorsal aorta (Plate 17, Figs. 3, 2), and finally it is cut frontally (Plate 16, Fig. 3). In more *anterior* sections it disappears.

Return to the level where the right postcardinal and common cardinal veins are broadly continuous (Plate 19, Fig. 1) and trace sections *posteriorly.* The postcardinal vein quickly separates from the common cardinal vein and lies lateral to the descending aorta (Plate 19, Fig. 3; Plate 20, Fig. 1; Plate 21, Fig. 1; Plate 22, Fig. 1). Try to identify small blood vessels continuous with the postcardinal vein dorsally; these vessels are the INTERSEGMENTAL VEINS. Note that the postcardinal vein lies above the developing mesonephric kidney (Plate 22, Fig. 1). Try to identify a small vessel lying beneath each mesonephric kidney; these vessels are the SUBCARDINAL VEINS. Trace the postcardinal vein caudally until it disappears.

Try to identify and trace the precardinal, common cardinal, and postcardinal veins on the *left* side of the embryo. These vessels are usually more difficult to identify with certainty than on the right side.

4. Serial sagittal sections. Aids: Plate 23; A, Fig. 576; HU, Fig. 137; M, Figs. 127-129.

Approximately the cranial half of the embryo is actually cut sagittally; the caudal half is cut frontally. As with the 48-hour stage find a section which is approximately a midsagittal section through the BRAIN and NOTOCHORD. Identify all regions of the brain including the INFUNDIBULUM and PINEAL GLAND. Also identify the PHARYNX, PREORAL GUT, STOMODEUM, RATHKE'S POUCH, and MOUTH OPENING. The mouth opening has formed by degeneration of the oral membrane. Following this degeneration the cavities of the foregut and stomodeum are continuous. If time permits try to identify as many other structures as possible.

5. Summary of germ layer derivatives formed since the 48-hour stage.

Ectoderm	*Mesoderm*	*Ectoderm and Somatic Mesoderm*
cerebral hemispheres	mesenteries:	leg buds
cranial ganglia:	dorsal mesogaster	wing buds
X. jugular	duodeno-hepatic ligament	
cranial nerves:	gastro-hepatic ligament	*Endoderm and Splanchnic Mesoderm*
III. oculomotor	mesoduodenum	allantois
V. mandibular, maxillary, and ophthalmic branches of trigeminal	mesoesophagus	
VII. facial	mesonephric ducts	*Ectoderm and Endoderm*
IX. glossopharyngeal	mesonephric tubules	cloacal membrane
endolymphatic ducts	*Endoderm*	
lens epithelium	cloaca	
lens fibers	dorsal pancreatic rudiment	
nasal pits	esophagus	
pineal gland	hindgut	
spinal ganglia	laryngo-tracheal groove	
	stomach	
	tail gut	

IV

10-mm Pig Embryos

A. WHOLE SPECIMEN

Beyond the 72-hour stage, the development of chick embryos becomes increasingly less typical of vertebrates in general and more avian in character. Consequently for more advanced development it is advisable to examine mammalian embryos. Among mammals pig embryos are preferred and are the most available in the stages and quantities needed. The 10-mm pig embryo, which has been developing for 20-21 days of a total gestation period of approximately 4 months (usual range 110-116 days), contains the rudiments of almost all adult structures. Therefore examination of this stage is almost a study of adult anatomy in miniature.

Examine models or, preferably, *whole specimens* preserved in 70% alcohol. Whole preserved specimens are far superior to any model and to specimens embedded in plastic. If preserved specimens are used, your instructor should remove the amnion and carefully cut the umbilical cord parallel to and as close to the body wall as possible (if these structures are still present). The preserved embryo should be examined in a small dish containing sufficient 70% alcohol to cover it; observations should be made with a dissecting microscope. Manipulate your embryo very carefully, preferably with the enlarged rounded end of a small glass rod. Make every effort not to damage the embryo so that it can be used repeatedly.

Note the general C-shape of the embryo. (See Fig. K; A, Fig. 598; M, Fig. 147; PA, Fig. 58.) This shape is due to the CRANIAL (level of MESENCEPHALON), CERVICAL (level of MYELENCEPHALON), and TAIL FLEXURES. Because of these pronounced flexures, the most *cranial* regions of your serially transverse sectioned embryo will not be seen in the most *anterior* sections of your set of slides. Likewise the most *caudal* regions of the embryo will not be found in the most *posterior* sections.

Identify the HEAD, TRUNK, TAIL, FORELEG and HINDLEG BUDS, EYES (note dark coloration which is due to PIGMENT GRANULES within the PIGMENTED RETINAS), NASAL PITS (note the raised ridges bounding them, the LATERAL and MEDIAL NASAL PROCESSES), and SOMITES. Also identify the regions of the PROSENCEPHALON, MESENCEPHALON, METENCEPHALON, MYELENCEPHALON (note its thin ROOF PLATE), and SPINAL CORD. The STOMODEUM partially subdivides the large FIRST BRANCHIAL ARCH on each side into MAXILLARY and MANDIBULAR PROCESSES. Immediately behind the mandibular processes are the large FIRST BRANCHIAL GROOVES, followed by the SECOND BRANCHIAL ARCHES, and then by the SECOND BRANCHIAL GROOVES. Partially hidden behind the second branchial arches and grooves are the THIRD BRANCHIAL ARCHES. They are directly in front of the small, deep CERVICAL SINUSES (combined THIRD and FOURTH BRANCHIAL GROOVES).

Within the relatively straight trunk the sites of three large developing organs can be seen. The most cranial organ, indicated by a large bulge, is the HEART. The bulge immediately behind the heart, towards which the foreleg buds project, indicates the LIVER. The depression between the heart and liver bulges indicates the location of the developing SEPTUM TRANSVERSUM

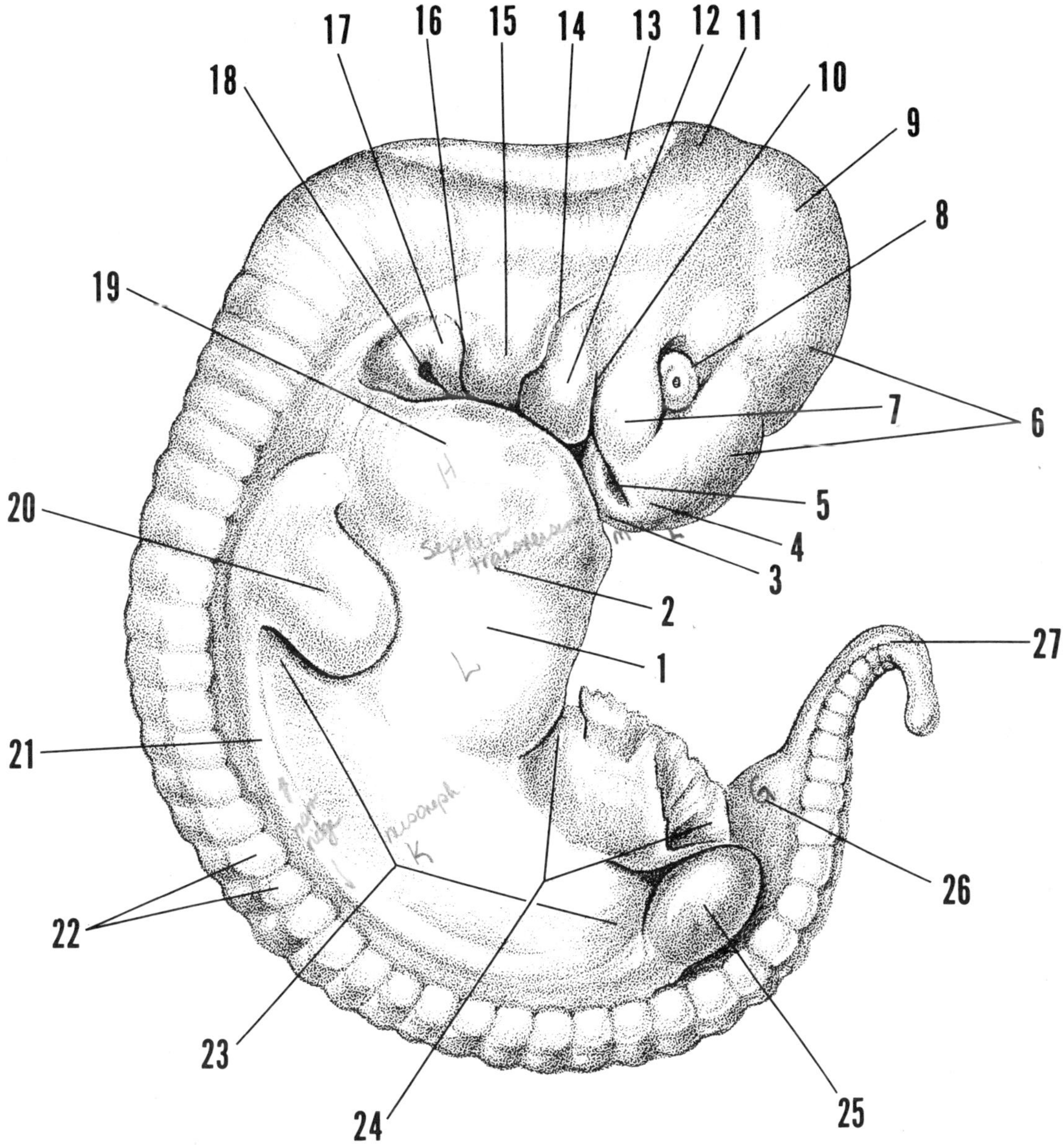

Fig. K. Drawing of the right side of a preserved 10-mm pig embryo.

1. Liver region
2. Location of developing septum transversum
3. Medial nasal process
4. Lateral nasal process
5. Nasal pit
6. Prosencephalon region
7. Maxillary process
8. Eye
9. Mesencephalon region
10. Stomodeum
11. Metencephalon region
12. Mandibular process
13. Myelencephalon roof plate region
14. First branchial groove
15. Second branchial arch
16. Second branchial groove
17. Third branchial arch
18. Cervical sinus
19. Heart region
20. Foreleg bud
21. Mammary ridge
22. Somites
23. Mesonephric kidney region
24. Umbilical cord
25. Hindleg bud
26. Genital eminence
27. Tail

(source of part of the adult DIAPHRAGM). The very large bulges, one on each side, between foreleg and hindleg buds, indicate the MESONEPHRIC KIDNEYS. Identify the MAMMARY RIDGES (small ridges, one on each side, lying parallel to the ventral edges of the somites between foreleg and hindleg buds) which form the MAMMARY GLANDS. The trunk ends with a ventral swelling, the GENITAL EMINENCE (rudiment of the EXTERNAL GENITALIA), which lies between the hindleg buds and the tapering tail. Any part of the UMBILICAL CORD still attached to the body is located between the liver region and hindleg buds.

B. SERIAL TRANSVERSE SECTIONS

Aids: Plates 24-37; M, Figs. 154-178; PA, Figs. 61-63, 69-79.

Position your slide on the microscope stage so that when viewed through the microscope each section is oriented as in Plate 24. Examine sections in antero-posterior sequence under low magnification unless otherwise directed. Since the amount of flexion encountered is quite variable and since the plane of section is not always exactly the same from embryo to embryo, the order of appearance of structures in *your* specimen will not always coincide *exactly* with the sequence described here, but should approximate it.

1. Nervous system. Aids: Fig. L; A, Figs. 599, 600; M, Fig. 149; PA, Figs. 59, 60, 92.

a. Brain. The first sections usually cut through the MYELENCEPHALON, but in some specimens the METENCEPHALON appears first. The former is recognized by its thin ROOF PLATE, the latter by its thick walls. The myelencephalon and metencephalon are usually cut frontally when first encountered. This is also true of many other cranial structures. The thin roof plate of the myelencephalon together with an outer adjacent vascular layer, the PIA MATER, are the rudiments of the CHOROID PLEXUS. (The pia mater is just developing and can be identified under high magnification as a thin layer of head mesenchymal cells just outside the brain wall. See Plate 24, Fig. 2.) The choroid plexus is formed by invagination of these rudiments into the cavity of the rhombencephalon, the FOURTH VENTRICLE, and is the source of CEREBROSPINAL FLUID (that is, the fluid which fills the cavities of the brain and spinal cord). Continue tracing sections posteriorly and note that another section of the brain appears near the pointed apex of the metencephalon; this is the MESENCEPHALON. The cavity of the mesencephalon, the CEREBRAL AQUEDUCT, soon becomes continuous with the fourth ventricle (Plate 24, Fig. 1). The constricted region between mesencephalon and metencephalon is the ISTHMUS.

The rhombencephalon and mesencephalon are the sources of several major adult brain components. The MYELENCEPHALON forms the MEDULLA; the METENCEPHALON forms the CEREBELLUM and PONS; and the MESENCEPHALON forms the CORPORA QUADRIGEMINA (SUPERIOR and INFERIOR COLLICULI) and the CEREBRAL PEDUNCLES.

Observe under high magnification that the walls of all regions of the brain are subdivided into three medio-lateral regions (Plate 24, Fig. 2). The innermost and darkest staining region (rich in nuclei) is the VENTRICULAR ZONE; the middle region containing fewer nuclei is the INTERMEDIATE (MANTLE) ZONE; and the outer region, relatively free of nuclei and the lightest staining, is the MARGINAL ZONE. The latter consists primarily of nerve fibers. The intermediate zone is formed by cells which migrate from the ventricular zone. At this stage most of these migratory cells are YOUNG NEURONS (formerly called NEUROBLASTS, a term no longer considered appropriate). Some migratory cells may be GLIOBLASTS (SPONGIOBLASTS), primitive nonnervous cells (that is, supporting cells) which remain within the wall of the brain.

Continue to trace sections posteriorly until the brain separates into two parts. The smaller oval-shaped (lower) section of the brain is still the mesencephalon; the larger (upper) section is still rhombencephalon (Plate 25, Figs. 1, 2). In more posterior sections the neural tube is separated into three parts (Plate 25, Fig. 3). The smaller (upper) section is either myelencephalon or SPINAL CORD (the boundary between myelencephalon and spinal cord is indistinct), the middle section is still rhombencephalon, and the remaining section is still mesencephalon. In more posterior sections the floor of the rhombencephalon fades out (Plate 26, Fig. 3). At about this same level the INFUNDIBULUM appears; it quickly becomes continuous with the DIENCEPHALON which now occupies the position occupied by the mesencephalon in more anterior sections. The cavity of the diencephalon is the major part of the THIRD VENTRICLE.

As you continue tracing sections posteriorly, identify RATHKE'S POUCH, a small flattened vesicle contacting the floor of the infundibulum (Plate 27, Figs. 2, 3). In more posterior sections it quickly becomes continuous with the STOMODEUM. At about this level the EYES can be identified. Quickly trace sections *anteriorly*. As the eyes fade out identify dense masses of mesodermal cells, the EYE MUSCLE RUDIMENTS, lying lateral to Rathke's pouch (Plate 27, Figs. 2, 3). These rudiments will later form the extrinsic EYE MUSCLES (INFERIOR and SUPERIOR OBLIQUE EYE MUSCLES; INFERIOR, SUPERIOR, INTERNAL, and EXTERNAL RECTUS EYE MUSCLES). Return to the level of the eyes and identify the following components (Plate 28): SENSORY and PIGMENTED RETINAS, LENS VESICLES, OPTIC STALKS, and OPTIC FISSURES. Observe under high magnification that the pigmented retinas now contain PIGMENT GRANULES. Each lens vesicle is completely separated from, and covered externally by, SKIN ECTODERM, the future CORNEAL EPITHELIUM. (The adult CORNEA is derived from both skin ectoderm and HEAD MESENCHYME.) Note that the head mesenchyme surrounding each pigmented retina is condensed, initiating formation of the CHOROID and SCLERA of the eye. Trace sections posteriorly and note that the lateral walls of the brain begin to bulge outwards to form the CEREBRAL HEMISPHERES of the TELENCEPHALON, and that the diencephalon fades out (Plate 29, Fig. 3; Plate 30, Fig. 1). The cavities of the cerebral hemispheres are the LATERAL VENTRICLES (or FIRST and SECOND VENTRICLES) and are broadly continuous with the cavity of the middle portion of the telencephalon (minor part of THIRD VENTRICLE). Later the connections between the lateral and third ventricles become greatly narrowed as the INTERVENTRICULAR FORAMINA. Continue tracing sections posteriorly and note that the two cerebral hemispheres separate from one another (Plate 31, Fig. 2). The portion of the telencephalon wall lying between the two cerebral hemispheres is the LAMINA TERMINALIS. Within the lamina terminalis COMMISSURES (transverse bundles of nerve fibers) will *later* form to interconnect the cerebral hemispheres.

The prosencephalon is the source of several major adult brain components. The DIENCEPHALON forms the THALAMUS, EPITHALAMUS, HYPOTHALAMUS, PINEAL GLAND, and POSTERIOR PITUITARY GLAND (derivative of the infundibulum); the TELENCEPHALON forms the CEREBRUM, OLFACTORY LOBES, and CORPORA STRIATA.

b. Cranial nerves. The cranial nerves will be described in order beginning with the first pair. Return to the most anterior section through the cerebral hemispheres (Plate 29, Fig. 3) and trace sections posteriorly. Identify the NASAL PITS. At first the cavity of each nasal pit is separated from the AMNIOTIC CAVITY, but these cavities become confluent in more posterior sections (Plate 30, Fig. 3). Many of the cells within the walls of the nasal pits are YOUNG NEURONS. Their AXONS are growing towards the cerebral hemispheres and may be identified in some specimens. On each side these axons constitute the OLFACTORY (I) CRANIAL NERVE (a SENSORY NERVE). DENDRITES will *later* develop from the young neurons within each nasal pit and differentiate as OLFACTORY RECEPTORS (receptors for the sense of smell). To either side of each nasal pit identify elevated regions of head mesenchyme covered by skin ectoderm, the LATERAL and MEDIAL NASAL PROCESSES. Each medial nasal process will eventually fuse laterally with a maxillary process and medially with the other medial nasal process to form the UPPER JAW. The lateral nasal processes eventually form the sides of the NOSE.

Return to the level of the optic fissures (Plate 28, Fig. 3) and examine the optic stalks and fissures under high magnification. Recall that within the region of each optic fissure the ventral lip of the optic cup is absent. Therefore at this level the MARGINAL ZONE of each sensory retina is continuous, via the marginal zone of the optic stalk, with the marginal zone of the floor of the diencephalon. Many of the cells within each sensory retina are YOUNG NEURONS. Some of these young neurons form GANGLION CELLS, whose AXONS grow into

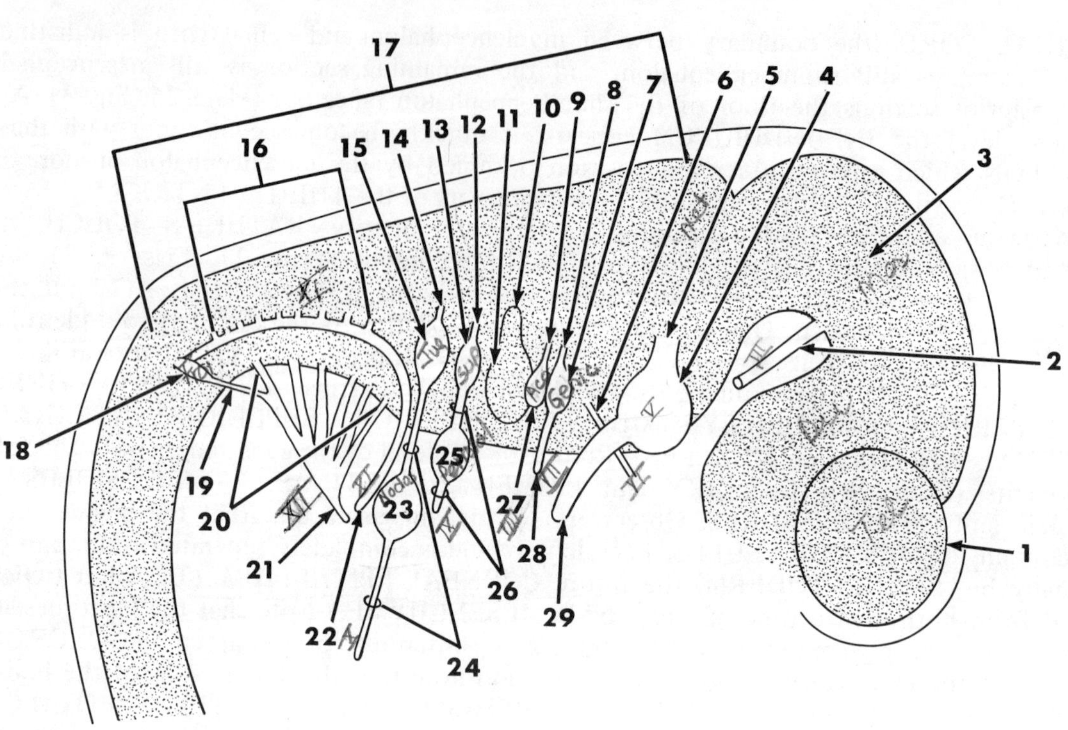

Fig. L. Schematic drawing of the cranial nerves and ganglia in the 10-mm pig embryo.

1. Cerebral hemisphere
2. Oculomotor nerve
3. Mesencephalon
4. Semilunar ganglion
5. Sensory root of the trigeminal nerve
6. Abducens nerve
7. Geniculate ganglion
8. Sensory root of the facial nerve
9. Auditory nerve
10. Endolymphatic duct
11. Auditory vesicle
12. Sensory root of the glossopharyngeal nerve
13. Superior ganglion
14. Sensory root of the vagus nerve
15. Jugular ganglion
16. Roots of the spinal accessory nerve
17. Rhombencephalon
18. Froriep's ganglion
19. Nerve fibers contributed by Froriep's ganglion to the hypoglossal nerve
20. Roots of the hypoglossal nerve
21. Hypoglossal nerve
22. Spinal accessory nerve
23. Nodosum ganglion
24. Vagus nerve
25. Petrosal ganglion
26. Glossopharyngeal nerve
27. Acoustic ganglion
28. Facial nerve
29. Mandibular branch of the trigeminal nerve

the marginal zone of the sensory retina. These axons remain within the marginal zone and grow to the floor of the diencephalon, constituting the OPTIC (II) CRANIAL NERVES (SENSORY NERVES). The optic nerves cannot be identified at this stage. (Since the axons constituting each optic nerve remain within the marginal zone throughout their extent, the so-called optic nerves are not actually true nerves, but rather are FIBER TRACTS located within derivatives of the diencephalon wall.)

Return to the level where the mesencephalon begins to separate from metencephalon (Plate 25, Fig. 1) and trace sections posteriorly. Notice two slight thickenings of the floor of the mesencephalon, one on either side of the midline. Each thickening is formed by an accumulation of CELL BODIES of YOUNG NEURONS within the INTERMEDIATE ZONE, and constitutes a MOTOR NUCLEUS. Identify AXONS growing out from these motor nuclei. On each side these axons constitute the OCULOMOTOR (III) CRANIAL NERVE (a MOTOR NERVE). Continue to trace sections posteriorly noting that the oculomotor nerves terminate near the eye muscle rudiments. These nerves will later innervate four pairs of extrinsic EYE MUSCLES (INFERIOR OBLIQUE EYE MUSCLES; INFERIOR, SUPERIOR, and INTERNAL RECTUS EYE MUSCLES).

The fourth cranial nerves, the TROCHLEAR CRANIAL NERVES (MOTOR NERVES), consist of AXONS which also grow out from MOTOR NUCLEI developing within the floor of the mesencephalon. (The trochlear nerves grow dorsad within the wall of the brain, cross over to the opposite side, and emerge from the brain in the isthmus region.) These nerves and their motor nuclei cannot be identified at this stage. In the adult the trochlear nerves innervate the SUPERIOR OBLIQUE EYE MUSCLES.

Return again to the level where mesencephalon begins to separate from metencephalon (Plate 25, Fig. 1). As sections are traced posteriorly, AXONS appear at the broadest part of the rhombencephalon, continuous with the latter (Plate 25, Fig. 2). On each side these axons constitute the SENSORY ROOT of the TRIGEMINAL (V) CRANIAL NERVE (a SENSORY and MOTOR, that is, MIXED NERVE). A few sections more posteriorly these sensory roots are continuous with the large SEMILUNAR GANGLIA (Plate 25, Fig. 3). Continue to trace sections posteriorly. As in the 72-hour chick embryo each semilunar ganglion eventually gives rise to three branches of the TRIGEMINAL NERVE: the OPHTHALMIC BRANCH (to eye region), MAXILLARY BRANCH (to maxillary process), and MANDIBULAR BRANCH (to mandibular process). In the 10-mm pig embryo usually only the mandibular branches can be readily identified (Plate 27, Fig. 1). The ophthalmic and maxillary branches are entirely sensory, consisting only of DENDRITES. The mandibular branches are eventually mixed, but at this stage consist only of DENDRITES. (*Later* AXONS will grow out from a pair of MOTOR NUCLEI within the floor of the rhombencephalon as the motor component of the mandibular branches.)

Return to the level where the neural tube is separated into three parts (Plate 25, Fig. 3) and trace sections posteriorly. As the rhombencephalon begins to fade out, identify AXONS growing out on each side from the floor of the rhombencephalon (Plate 26, Fig. 2). These axons constitute the ABDUCENS (VI) CRANIAL NERVES (MOTOR NERVES). The abducens nerves are usually cut across frontally, are very small, and quickly disappear as sections are traced posteriorly. They will later innervate the EXTERNAL RECTUS EYE MUSCLES.

Return to a level similar to Plate 24, Fig. 1 and trace sections posteriorly. Note that a small vesicle soon appears on each side of the myelencephalon; these vesicles are the ENDOLYMPHATIC DUCTS. In slightly more posterior sections identify the AUDITORY VESICLES which lie lateral to the endolymphatic ducts (Plate 24, Fig. 3). The dorsal portion of each auditory vesicle forms the UTRICULUS of the INNER EAR. In more posterior sections the endolymphatic duct and auditory vesicle on each side are continuous (Plate 25, Fig. 1). Continue to trace sections posteriorly and note that each auditory vesicle contains two smaller projections (Plate 25, Fig. 2). The projection nearer the top of the section is the rudiment of the POSTERIOR SEMICIRCULAR CANAL. The projection nearer the bottom is the rudiment of the ANTERIOR SEMICIRCULAR CANAL. The rudiments of the anterior and posterior semicircular canals are attached to the dorsal portion of the auditory vesicle (that is, the future utriculus). (*Later* the rudiment of the LATERAL SEMICIRCULAR CANAL will also form from the utriculus portion of each auditory vesicle.) In more posterior sections the rudiments of the anterior and posterior semicircular canals disappear; this portion of each auditory vesicle is the future SACCULUS of the INNER EAR (Plate 25, Fig. 3). The utriculi, semicircular canals, and sacculi form receptors for maintaining equilibrium. As sections are traced posteriorly the auditory vesicles narrow and fade out. The ventral portion of each auditory vesicle becomes the COCHLEA of the INNER EAR (Plate 26, Fig. 2), which forms the receptor organ for sound (that is, the SPIRAL ORGAN OF CORTI).

The next two cranial nerves are closely associated with each other and with the auditory vesicles and will therefore be described together. Return to a level similar to Plate 25, Fig. 3 and note that two ganglia lie just below each future sacculus. The lowermost ganglion is the GENICULATE GANGLION of the FACIAL (VII) CRANIAL NERVE (a MIXED NERVE); the uppermost ganglion is the ACOUSTIC GANGLION of the AUDITORY (VIII) CRANIAL NERVE (a SENSORY NERVE). At about this level, or in slightly more posterior sections, AXONS constituting the SENSORY ROOTS of the FACIAL NERVES interconnect the geniculate ganglia and rhombencephalon (Plate 26, Fig. 1). (AXONS constituting the AUDITORY NERVES likewise interconnect the acoustic ganglia and rhombencephalon, but since the auditory nerves are very short and closely applied to the sensory roots of the facial nerves they are difficult to identify with certainty.) Continue to trace sections posteriorly and identify the FACIAL NERVES extending from the geniculate ganglia into the *second* branchial arches (Plate 26, Fig. 2). At this stage the facial nerves consist only of DENDRITES. (*Later* AXONS will grow out from a pair of MOTOR NUCLEI within the floor of the rhombencephalon as the motor component of the facial nerves.) Note that the acoustic ganglia and auditory vesicles fade out at about the same level (Plate 26, Fig. 2). (Each acoustic ganglion eventually subdivides into two parts [usually at later stages], the VESTIBULAR and SPIRAL [COCHLEAR] GANGLIA. DENDRITES from CELL BODIES of YOUNG NEURONS within the VESTIBULAR GANGLIA terminate in equilibrium receptors in the semicircular canals, utriculi, and sacculi; DENDRITES from CELL BODIES of YOUNG NEURONS within the SPIRAL GANGLIA terminate in the spiral organ of Corti within each cochlea.)

Return to the level at which the auditory vesicle and endolymphatic duct on each side become continuous (Plate 25, Fig. 1). Identify AXONS continuous with the wall of the myelencephalon just above each auditory vesicle. On each side these axons constitute the SENSORY ROOT of the GLOSSOPHARYNGEAL (IX) CRANIAL NERVE (a MIXED NERVE). In slightly more posterior sections the SUPERIOR GANGLIA of the glossopharyngeal nerves appear (Plate 25, Fig. 2). CELL BODIES of YOUNG NEURONS within these ganglia form both the AXONS interconnecting the superior ganglia and myelencephalon, and the DENDRITES which constitute the part of the GLOSSOPHARYNGEAL NERVES appearing in place of the superior ganglia in slightly more posterior sections (Plate 25, Fig. 3). Continue tracing the glossopharyngeal nerves posteriorly. Note that another pair of ganglia (much larger than the superior ganglia) appears in place of the glossopharyngeal nerves (Plate 26, Fig. 2). These are the PETROSAL GANGLIA of the glossopharyngeal nerves. CELL BODIES of YOUNG NEURONS within these ganglia also form AXONS (impossible to identify) interconnecting the petrosal ganglia and myelencephalon, and DENDRITES, which along with dendrites growing out from the superior ganglia constitute the part of the GLOSSOPHARYNGEAL NERVES appearing in place of the petrosal ganglia in slightly more posterior sections (Plate 27, Figs. 1, 2). (*Later* AXONS constituting the motor component of the glossopharyngeal nerves will grow out from a pair of MOTOR NUCLEI within the floor of the myelencephalon.) Try to trace the glossopharyngeal nerves into the *third* branchial arches where they terminate.

Return to a section similar to Plate 24, Fig. 1 and trace sections posteriorly. At about the level at which the endolymphatic ducts appear, identify AXONS continuous with the wall of the myelencephalon just caudal to the auditory vesicles (Plate 24, Fig. 3). On each side these axons constitute the SENSORY ROOT of the VAGUS (X) CRANIAL NERVE (a MIXED NERVE). A few sections more posteriorly the large JUGULAR GANGLIA of the vagus nerves appear (Plate 25, Fig. 1). Note that the jugular ganglia lie just caudal to the superior ganglia. In more posterior sections the jugular ganglia fade out and the VAGUS NERVES appear in their place (Plate 25, Fig. 3). (This part of each vagus nerve consists of DENDRITES derived from CELL BODIES of YOUNG NEURONS within each jugular ganglion.) Continue tracing sections posteriorly and observe that the vagus nerves fade out and that the NODOSUM GANGLIA (derived from a pair of EPIBRANCHIAL PLACODES) appear in their place (Plate 27, Fig. 2). In more posterior sections the nodosum ganglia fade out in the area of the *fourth* branchial arches. It is difficult to trace the VAGUS NERVES caudal to this level. In many embryos they have extended caudad and lie ventro-lateral to the esophagus (Plate 31, Fig. 2). At these caudal levels the vagus nerves consist primarily of PREGANGLIONIC AXONS of the PARASYMPATHETIC DIVISION of the AUTONOMIC NERVOUS SYSTEM. These axons grow out from MOTOR NUCLEI (impossible to identify) developing within the floor of the myelencephalon.

Return to a level similar to Plate 24, Fig. 1 and trace sections posteriorly. Identify AXONS growing out from the myelencephalon (Plate 24, Fig. 2). On each side these axons constitute the ROOTS of the SPINAL ACCESSORY (XI) CRANIAL NERVE (a MOTOR NERVE). A few sections more posteriorly these roots merge together on each side and are cut frontally as the SPINAL ACCESSORY NERVE (Plate 24, Fig. 3). A few sections more posteriorly each spinal accessory nerve is cut across twice (Plate 25, Fig. 1). The part of each spinal accessory nerve nearest the top of the section can be traced caudally where it eventually fades out adjacent to the spinal cord. Trace the lower part of each spinal accessory nerve caudally. It first lies adjacent to the jugular ganglion, then adjacent to the vagus nerve (Plate 25, Fig. 3), then adjacent to the nodosum ganglion (Plate 27, Fig. 3), and finally it fades out. In the adult the spinal accessory nerves primarily innervate derivatives of the *fourth* branchial arches and certain NECK and SHOULDER MUSCLES.

Return to the level where the spinal accessory nerves are cut frontally (Plate 24, Fig. 3) and trace sections posteriorly. Identify several bundles of AXONS growing out from the myelencephalon (Plate 25, Fig. 2). On each side these bundles constitute the ROOTS of the HYPOGLOSSAL (XII) CRANIAL NERVE (a MOTOR NERVE). Trace these roots as far caudally as possible. In some embryos it may be possible to observe that on each side the roots merge together and continue caudally as the HYPOGLOSSAL NERVE. In the adult the hypoglossal nerves innervate the TONGUE MUSCLES.

The following chart summarizes the development of cranial nerves through the 10-mm stage.

Cranial Nerves	*Cranial Ganglia*	*Origin of Cranial Ganglia*	*Type of Nerve Fibers Present*	*Regions Innervated*
olfactory	———	———	sensory	lining of nasal cavities
optic (nerves cannot be identified)	———	———	sensory	retinas
oculomotor	———	———	motor	inferior oblique eye muscles; inferior, superior, and internal rectus eye muscles
trochlear (nerves cannot be identified)	———	———	motor	superior oblique eye muscles
trigeminal	semilunar	neural crest cells and epibranchial placodes	sensory (*later* also motor)	primarily derivatives of first branchial arches
abducens	———	———	motor	external rectus eye muscles
facial	geniculate	neural crest cells and epibranchial placodes	sensory (*later* also motor)	primarily derivatives of second branchial arches
auditory	acoustic	auditory placodes	sensory	inner ears
glossopharyng-eal	superior and petrosal	neural crest cells and epibranchial placodes, respectively	sensory (*later* also motor)	primarily derivatives of third branchial arches
vagus	jugular and nodosum	neural crest cells and epibranchial placodes, respectively	sensory and motor	primarily derivatives of fourth branchial arches
spinal accessory	———	———	motor	primarily derivatives of fourth branchial arches and certain neck and shoulder muscles
hypoglossal	———	———	motor	tongue muscles

c. ***Spinal cord and spinal nerves.*** Return to the level where the spinal accessory nerves are cut frontally (Plate 24, Fig. 3) and trace sections posteriorly, watching the uppermost part of the neural tube in your sections. On each side observe a rather small but prominent ganglion just caudal to the roots of the spinal accessory nerve; this is FRORIEP'S GANGLION (Plate 25, Fig. 2) which contributes nerve fibers to the hypoglossal nerve. The level of Froriep's ganglia is still the level of the myelencephalon, but a few sections more posteriorly the SPINAL CORD gradually appears. At about this level identify a pair of ganglia lying adjacent to the cranial end of the spinal cord, the most cranial SPINAL GANGLIA.

Quickly trace the spinal cord caudally and find one or more sections which show all of the following structures (see Plate 34, Fig. 2; A, Figs. 470, 618-623): DORSAL ROOT (composed of AXONS entering the spinal cord, SPINAL GANGLION, and DENDRITES entering the spinal ganglion), VENTRAL ROOT (composed of AXONS leaving the spinal cord), and SPINAL NERVE with its three branches, the DORSAL RAMUS, VENTRAL RAMUS, and VISCERAL RAMUS (RAMUS COMMUNICANS). The visceral ramus on each side extends ventro-medially towards an accumulation of NEURAL CREST CELLS, one of the SYMPATHETIC CHAIN GANGLIA of the AUTONOMIC NERVOUS SYSTEM. (The SYMPATHETIC COLLATERAL and PARASYMPATHETIC TERMINAL GANGLIA of the AUTONOMIC NERVOUS SYSTEM will *later* form from other accumulations of NEURAL CREST CELLS.)

Identify the following components of the spinal cord: VENTRICULAR, INTERMEDIATE, and MARGINAL ZONES, ALAR PLATE (dorsal half of spinal cord), BASAL PLATE (ventral half of spinal cord), SULCUS LIMITANS (slight depression on each side partially separating alar and basal plates), ROOF and FLOOR PLATES, and MOTOR HORNS (paired enlargements of the ventral part of the intermediate zone). Note the small diameter of the NOTOCHORD; it is enclosed within a mass of sclerotome cells which will later form the CENTRUM of each VERTEBRA. Most of the notochord later degenerates.

At the level of the foreleg buds, the ventral rami of adjacent spinal nerves are interconnected in complex ways forming the BRACHIAL PLEXUS (Plate 32, Fig. 1). (*Later* a LUMBOSACRAL PLEXUS will form at the level of the hindleg buds.) Note that the FORELEG and HINDLEG BUDS consist of an outer layer of ectoderm, part of which is thickened as the APICAL ECTODERMAL RIDGE, and a core of somatic mesoderm. At section levels between the limb buds note a small dorso-lateral swelling on each side, the MAMMARY RIDGE (Plate 32, Fig. 3).

Trace the spinal cord caudally until it is cut frontally due to the tail flexure. Identify the DERMATOME, MYOTOME, and SCLEROTOME subdivisions of the SOMITES. Note that each sclerotome has a dense *caudal* portion and a less dense *cranial* portion. Each VERTEBRA is formed by the fusion of the caudal half of one sclerotome with the cranial half of the next caudal sclerotome. By examining sections cranial or caudal to this level, identify the SPINAL GANGLIA and the VENTRAL ROOTS of the SPINAL NERVES, which leave the ventro-lateral walls of the spinal cord.

2. **Respiratory and digestive systems. Aids:** Fig. M; A, Figs. 601-604; M, Fig. 149; PA, Figs. 60, 64.

Find the most anterior section showing a round ESOPHAGUS (Plate 30, Fig. 2). The round TRACHEA usually lies beneath the esophagus at this level. Trace sections posteriorly a considerable distance watching both esophagus and trachea. Note a distinct asymmetrical evagination of the trachea towards the *right* (apparent left) side. This evagination is the EPARTERIAL BRONCHUS (Plate 31, Fig. 2); it will form the upper lobe of the right LUNG. Continue tracing sections posteriorly and observe that a vesicle appears on each side of the trachea and quickly becomes continuous with the latter. These vesicles are the LUNG BUDS. In more posterior sections the trachea fades out and the lung buds continue caudally for a short distance (Plate 32, Fig. 1). (Each of the two lung buds *later* bifurcates to form two lobes of the lung. Therefore in the adult the *right* lung contains three lobes, the *left* lung, only two.) At this level identify the PLEURAL CAVITIES lying lateral to the lung buds, and note that they are continuous laterally with the PERITONEAL CAVITY. A thin layer of splanchnic mesoderm, the VISCERAL PLEURA, lies just medial to each pleural cavity (examine this layer under high magnification). The PARIETAL PLEURA has not yet formed at this level. More cranially the pleural cavities are isolated from all other portions of the coelom, and visceral and parietal pleura can be identified (Plate 31, Fig. 3).

Return to a level similar to Plate 32, Fig. 1 and trace sections posteriorly, noting that the esophagus shifts ventrad and the lung buds fade out (Plate 32, Fig. 2). Identify the MESO-ESOPHAGUS. In more posterior sections the STOMACH appears (Plate 32, Fig. 3). The stomach has undergone a rotation around its longitudinal axis of approximately 45° so that its original *dorsal* surface lies to the *left* (apparent right) and its *ventral* surface lies to the *right* (apparent left). The original dorsal surface forms the GREATER CURVATURE of the adult stomach; the original ventral surface forms the LESSER CURVATURE. Identify the DORSAL MESO-GASTER, which enlarges greatly as the adult GREATER OMENTUM. (*Later* mesenchymal cells will aggregate within the dorsal mesogaster and form the SPLEEN.) Also identify the mesentery

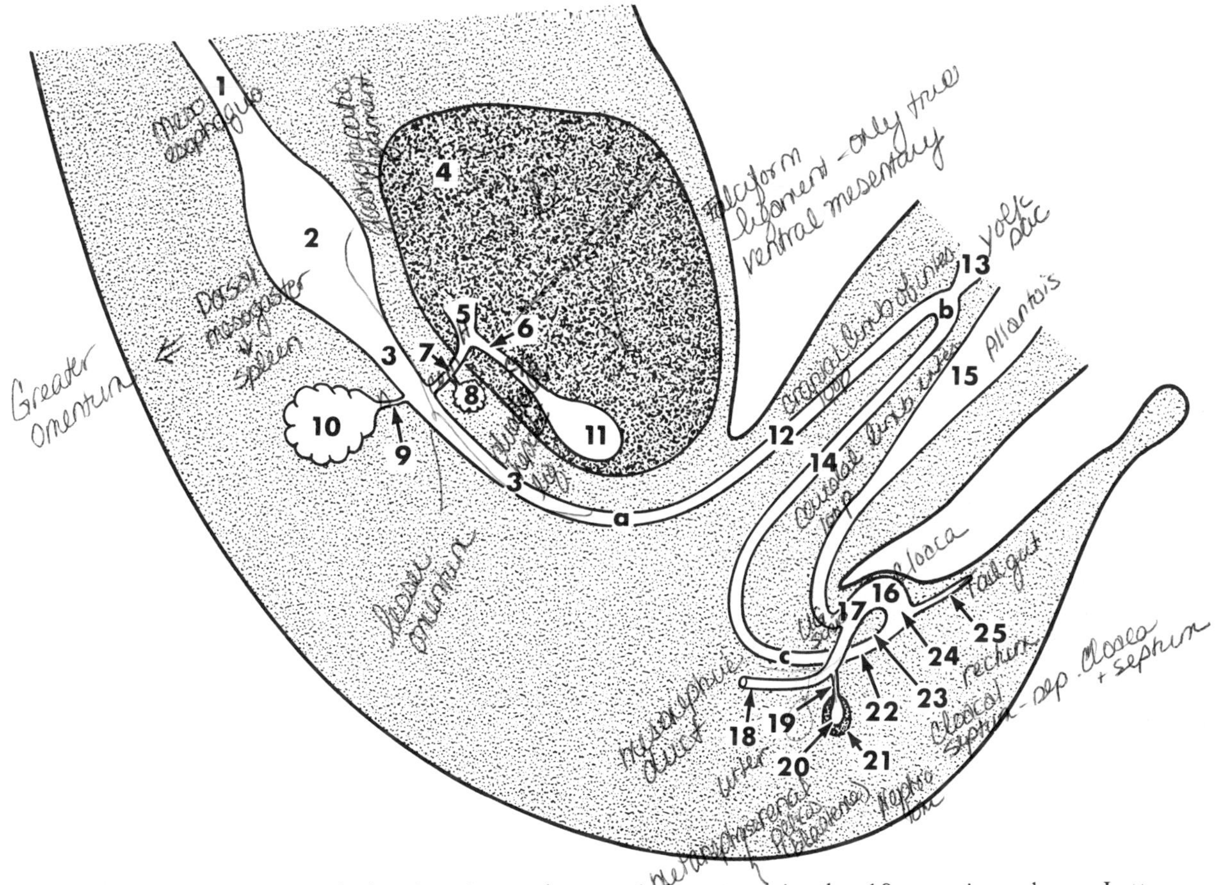

Fig. M. Schematic drawing of the digestive and urogenital systems in the 10-mm pig embryo. Letters a-c indicate gut levels described in text. The outer line indicates the boundaries of the caudal two-thirds of the embryo and of the umbilical cord.

1. Esophagus
2. Stomach
3. Duodenum
4. Liver
5. Hepatic duct
6. Cystic duct
7. Common bile duct
8. Ventral pancreatic rudiment
9. Dorsal pancreatic duct
10. Dorsal pancreatic rudiment
11. Gall bladder
12. Cranial limb of the intestinal loop
13. Yolk sac
14. Caudal limb of the intestinal loop
15. Allantois
16. Cloaca
17. Urogenital sinus
18. Mesonephric duct
19. Ureter
20. Renal pelvis
21. Nephrotome
22. Colon
23. Cloacal septum
24. Rectum
25. Tail gut

connecting stomach and LIVER, the GASTRO-HEPATIC LIGAMENT. Note that the liver is subdivided into paired DORSAL and VENTRAL LOBES and that it consists of irregularly shaped cords of cells, the HEPATIC CORDS, separated by irregular vascular spaces, the HEPATIC SINUSOIDS. At this stage the liver is functioning as a HEMOPOIETIC ORGAN (that is, it is temporarily producing blood cells). Identify the OMENTAL BURSA, a closed cavity to the *right* (apparent left) of the stomach. Trace sections posteriorly, noting that the omental bursa opens into the PERITONEAL CAVITY via the EPIPLOIC FORAMEN (Plate 33, Fig. 2). Lining the peritoneal cavity are the VISCERAL PERITONEUM (a thin layer of splanchnic mesoderm on the surface of the abdominal organs) and the PARIETAL PERITONEUM (a thin layer of somatic mesoderm on the internal surface of the ventral and lateral body walls); examine these layers under high magnification.

Return to the level of the stomach (Plate 32, Fig. 3) and trace sections posteriorly, noting that the stomach fades out and that the DUODENUM appears in its place (Plate 33, Fig. 1). At about this level sections also cut through the endodermal HEPATIC DUCT, located within the dorsal part of the LIVER. The mesentery connecting duodenum and liver is the DUODENO-HEPATIC LIGAMENT. (The gastro-hepatic and duodeno-hepatic ligaments collectively constitute the adult LESSER OMENTUM.) The mesentery connecting the liver to the ventral body wall is the FALCIFORM LIGAMENT. In slightly more posterior sections the hepatic duct constricts into an upper COMMON BILE DUCT, which quickly joins the duodenum, and a lower CYSTIC DUCT (Plate 33, Fig. 2). In more posterior sections the cystic duct gradually expands as the GALL BLADDER. In some specimens the gall bladder may be bilobed or trilobed. Continue to trace sections posteriorly until the liver disappears. Then reverse direction and trace sections *anteriorly,* noting the reappearance of the liver and finally the disappearance of its cranial end. Note that the cranial end of the liver terminates in a transverse partition, the SEPTUM TRANSVERSUM (Plate 32, Fig. 1).

Return to the level where the common bile duct joins the duodenum (Plate 33, Fig. 2) and trace sections posteriorly. As the common bile duct disappears, a small darkly stained solid mass of cells, the VENTRAL PANCREATIC RUDIMENT, appears in its place (Plate 33, Fig. 3). The ventral pancreatic rudiment develops as an outgrowth from the common bile duct; its original connection to the common bile duct, the VENTRAL PANCREATIC DUCT (DUCT OF WIRSUNG), is degenerating and probably cannot be identified. (In humans this connection usually persists.) At about this same level identify the DORSAL PANCREATIC RUDIMENT. This rudiment develops as an outgrowth from the duodenum; its connection to the duodenum, the DORSAL PANCREATIC DUCT (DUCT OF SANTORINI), persists in the adult pig to transport certain pancreatic secretions involved in digestion to the duodenum. (In humans this connection usually degenerates.) Later both dorsal and ventral pancreatic rudiments will fuse to form the PANCREAS.

Trace sections posteriorly from the level at which the dorsal pancreatic duct connects to the duodenum (Plate 33, Fig. 3) and note that the duodenum lies dorsal to another region of the gut contained within the lightly stained tissue of the UMBILICAL CORD (that is, contained within WHARTON'S JELLY, a mesodermal derivative). This latter region of the gut is the CRANIAL LIMB OF THE INTESTINAL LOOP (Plate 34, Fig. 1). The INTESTINAL LOOP has herniated into the EXTRA-EMBRYONIC COELOM of the umbilical cord forming the temporary UMBILICAL HERNIA. (The cranial limb of the intestinal loop later forms primarily the JEJUNUM and most of the ILEUM of the adult SMALL INTESTINE.) As sections are traced posteriorly the duodenum and cranial limb of the intestinal loop progressively approach one another and become continuous. At this level these regions of the gut are cut frontally (Fig. M, level a; Plate 35, Fig. 1), and they quickly disappear as sections are traced posteriorly. Return to the level where the duodenum and cranial limb of the intestinal loop are cut frontally and trace sections *anteriorly* observing the cranial limb of the intestinal loop. Note that it rotates somewhat towards the *right* (apparent left) to lie alongside another portion of the gut, the CAUDAL LIMB OF THE INTESTINAL LOOP (Plate 34, Fig. 3). (The caudal limb of the intestinal loop later forms primarily the ASCENDING COLON and most of the TRANSVERSE COLON of the adult LARGE INTESTINE.) Trace sections *anteriorly* a considerable distance and try to observe the continuity of the two limbs at the apex of the intestinal loop (see Fig. M, level b; A, Figs. 601, 603). The YOLK SAC normally attaches to the intestinal loop at its apex. If the yolk sac is present in your specimen, it will first appear as a slight circular expansion of the gut tube as the apex of the intestinal loop is traced *anteriorly*. In more *anterior* sections the yolk sac appears as a large, irregularly shaped vesicle (Plate 32, Fig. 2). (Unfortunately, in some specimens the umbilical cord has been cut off so close to the body that the apex of the intestinal loop and the yolk sac are not present in your sections.)

Return to the level of the apex of the intestinal loop, or as close to this level as possible, and trace sections *posteriorly*, observing the caudal limb of the intestinal loop. It rotates somewhat towards the *left* (apparent right) and is eventually suspended between the two large mesonephric kidneys by a long, thin mesentery (Plate 35, Fig. 3). Since this portion of the caudal limb of the intestinal loop will form part of the COLON of the adult LARGE INTESTINE, this mesentery is part of the MESOCOLON. As you continue to trace sections posteriorly, the caudal limb of the intestinal loop approaches another region of the gut. This latter region is the caudal part of the COLON (Plate 36, Fig. 2); it quickly joins the caudal limb of the intestinal loop, and these two structures are cut frontally (Fig. M, level c). They then quickly disappear as sections are traced posteriorly. Reverse direction and trace sections *anteriorly* following the colon (that is, the lower part of the gut in your sections). Note that it becomes slightly elongated dorso-ventrally as the RECTUM, which then quickly joins a large cavity, the CLOACA (Plate 35, Fig. 3). At this stage the cloaca is subdividing into a dorsal (lower) RECTUM and a ventral (upper) UROGENITAL SINUS. The mesodermal ingrowth which separates the cloaca into these two chambers is the CLOACAL SEPTUM.

Trace sections *anteriorly* while watching the cloaca, and note that as it fades out a solid cord of cells, the CLOACAL MEMBRANE, can be identified (Plate 35, Fig. 2). (This membrane appears different than that seen in the 72-hour chick embryo since it is cut frontally rather than transversely.) Note that the cloacal membrane is contained within a large swelling, the GENITAL EMINENCE. In some specimens a small, degenerating TAIL GUT can be identified at about this level.

Trace sections *posteriorly* and try to observe that the tail gut opens cranially into the cloaca. Note that as sections are traced posteriorly, the cloaca becomes completely subdivided into the UROGENITAL SINUS and RECTUM. In more posterior sections the COLON again appears (note its rounded rather than elongated shape) and an endoderm-lined cavity, the ALLANTOIS, emerges from the umbilical cord to join the urogenital sinus ventrally (Plate 36, Fig. 1). Reverse direction and trace sections *anteriorly*, following the allantois out into the umbilical cord (Plates 35, 34). Note that within the umbilical cord the allantois lies between two large arteries, the UMBILICAL (ALLANTOIC) ARTERIES.

Return to the most anterior section showing the round TRACHEA beneath the round ESOPHAGUS (Plate 30, Fig. 2) and trace sections *anteriorly*. The trachea elongates dorso-ventrally as the LARYNX which lies beneath the somewhat triangular-shaped esophagus. (The walls of the larynx tend to be in contact and may therefore be difficult to identify.) In more *anterior* sections the esophagus expands as the PHARYNX and the latter immediately becomes continuous with the larynx (Plate 29, Fig. 2). The opening of the larynx into the pharynx is the GLOTTIS. At about this level the FOURTH PHARYNGEAL POUCHES can be identified. They usually first appear as isolated vesicles ventro-lateral to the pharynx, but then quickly join the pharynx as sections are traced *anteriorly*. The fourth pharyngeal pouches are the sources of the adult SUPERIOR PARATHYROID GLANDS. In some specimens a small dorso-lateral evagination of each fourth pharyngeal pouch can be identified (Plate 29, Fig. 1). These are the rudimentary FIFTH PHARYNGEAL POUCHES (ULTIMOBRANCHIAL BODIES). They are later incorporated into the thyroid gland where they give rise to C (PARAFOLLICULAR) CELLS, which produce the hormone CALCITONIN.

As you continue to trace sections *anteriorly*, note a broad ectodermal invagination on each side lateral to the pharynx (Plate 28, Fig. 2), the CERVICAL SINUS (combined THIRD and FOURTH BRANCHIAL GROOVES). At about this level the larynx usually opens broadly into the pharynx via the glottis. Note the prominent swellings on either side of the glottis, the

ARYTENOID SWELLINGS. These will later form the ARYTENOID CARTILAGES of the LARYNX. At about this level identify the THYROID GLAND, lying beneath the floor of the pharynx. Its original connection to the floor of the pharynx, the THYROGLOSSAL DUCT, has degenerated. Such a ductless gland which drains its secretions only into the blood stream is an ENDOCRINE GLAND. Continue to trace sections *anteriorly* and identify a midline swelling of the floor of the pharynx just cranial to the glottis. This swelling is the EPIGLOTTIS (Plate 28, Fig. 1). It is derived from the third and fourth branchial arches.

Return again to the most posterior section through the glottis (Plate 29, Fig. 2) and trace sections *anteriorly.* Identify a pair of flattened vesicles lying just above (dorsal to) the third branchial arches, the THIRD PHARYNGEAL POUCHES (Plate 28, Fig. 3). The ventral portions of the two third pharyngeal pouches (that is, the portions shown in Plate 28, Fig. 3) *later* fuse together in the midline to form the THYMUS GLAND. As sections are traced *anteriorly* the third pharyngeal pouches move dorsally and become continuous with the pharynx. Identify the THIRD CLOSING PLATES at about this level. The dorsal portion of each third pharyngeal pouch (that is, the portion shown in Plate 28, Figs. 1, 2) forms the INFERIOR PARATHYROID GLANDS. The rudiments of these glands can be seen in most embryos under high magnification as a thickening just beneath the cavity of the third pharyngeal pouch on each side (see A, Fig. 613). (Later the entire third pharyngeal pouches, including the rudiments of the inferior parathyroid and thymus glands, migrate caudal to the fourth and fifth pharyngeal pouches. Thus in the adult the inferior parathyroid glands [derived from the third pharyngeal pouches] lie *caudal* to the superior parathyroid glands [derived from the fourth pharyngeal pouches]).

Continue to trace sections *anteriorly* and identify the SECOND PHARYNGEAL POUCHES (Plate 27, Figs. 3-1), the sources of the adult PALATINE TONSILS. Also identify the SECOND BRANCHIAL GROOVES and SECOND CLOSING PLATES. The latter may be ruptured in part forming the SECOND BRANCHIAL CLEFTS. If the closing plates are partially ruptured the exact boundary between the second branchial groove and pharyngeal pouch on each side may be difficult to determine. At about the level of the second pharyngeal pouches identify the TONGUE RUDIMENTS. These consist of a pair of LATERAL LINGUAL SWELLINGS, a midline TUBERCULUM IMPAR, and a midline COPULA (Plate 27, Figs. 2, 3). The copula is derived from the second branchial arches; the lateral lingual swellings and tuberculum impar from the first branchial arches (although these three latter structures are now fused with the second branchial arches). Note that the tongue rudiments protrude into the pharynx.

Trace sections *anteriorly.* At about the level at which the tongue rudiments fade out identify the FIRST BRANCHIAL GROOVES, CLOSING PLATES, and PHARYNGEAL POUCHES (Plate 27, Fig. 1; Plate 26, Fig. 3). On each side the first branchial groove forms the adult EXTERNAL AUDITORY MEATUS (EXTERNAL EAR CANAL). Each first closing plate, after mesodermal cells penetrate between its ectoderm and endoderm, forms a TYMPANIC MEMBRANE. Note that each first pharyngeal pouch consists of two portions (Plate 27, Fig. 1). The upper portion in your sections connects to the pharynx and forms the adult EUSTACHIAN (PHARYNGO-TYMPANIC) TUBE; the lower portion forms the TYMPANIC (MIDDLE EAR) CAVITY. On each side the tympanic cavity will eventually surround the EAR OSSICLES (that is, the MALLEUS and INCUS [derivatives of the first branchial arch] and the STAPES [derivative of the second branchial arch]). Elevations will *later* form from the first and second branchial arches to either side of the first branchial grooves. These elevations will subsequently fuse on each side to form the PINNA (AURICLE) of the EXTERNAL EAR.

3. Urogenital system. Aids: Fig. M; A, Fig. 601, 603; M, Fig. 149; PA, Figs. 60, 64.

Return to the level of the cloacal membrane (Plate 35, Fig. 2) and trace sections posteriorly until the cloaca is reached. Note that the ventral (upper) portion of the cloaca broadens transversely. This broadened portion is the UROGENITAL SINUS. In more posterior sections the urogenital sinus is continuous with the MESONEPHRIC DUCTS (Fig. M; Plate 36, Fig. 1).

Continue tracing sections posteriorly following the mesonephric ducts. Note that they lie lateral to the colon (Plate 36, Fig. 2). They are quickly cut frontally as they join another portion of the mesonephric ducts located in the ventral part of the large MESONEPHRIC KIDNEYS (Plate 36, Fig. 3). A few sections more posteriorly the mesonephric ducts disappear.

Return to the level where the mesonephric ducts are cut frontally (Plate 36, Fig. 3) and trace that part of each mesonephric duct within the mesonephric kidney *anteriorly*. Note that numerous MESONEPHRIC TUBULES are continuous with it. Continue to trace sections *anteriorly* until the mesonephric ducts can no longer be identified. The mesonephric duct on each side forms the EPIDIDYMIS, VAS DEFERENS, and EJACULATORY DUCT of the adult male reproductive system and also gives rise to an evagination which forms the SEMINAL VESICLE. The mesonephric ducts primarily degenerate in the female.

Note the tremendous size of the mesonephric kidneys relative to that of other structures of the pig embryo (see A, Fig. 599; M, Fig. 148; PA, Fig. 59). This large size probably counteracts the rather inefficient placenta of pig embryos (DIFFUSE, EPITHELIO-CHORIALIS TYPE with a PLACENTAL MEMBRANE [BARRIER] composed of many layers). Apparently the placenta does not remove nitrogenous wastes from the blood stream of pig embryos very readily; pig embryos therefore possess massive mesonephric kidneys to take care of this function for themselves. Examine a section cut through about the middle of the mesonephric kidneys (Plate 35). Note that this type of kidney consists mostly of MESONEPHRIC TUBULES. These tubules primarily degenerate in females. In males some mesonephric tubules persist as the EFFERENT DUCTULES (VASA EFFERENTIA) of the adult reproductive system; the remaining tubules degenerate. At the medial side of each mesonephric kidney note several large spaces bounded by a very flat epithelium and filled with cells. These expanded spaces are the GLOMERULAR (BOWMAN'S) CAPSULES of the mesonephric tubules (Plate 35, Fig. 1); these structures later degenerate in both males and females. The cells contained within the glomerular capsules are capillaries which constitute the GLOMERULI, and blood cells contained within the glomeruli (Plate 35, Fig. 2). These glomeruli likewise later degenerate in both males and females.

The GONAD RUDIMENTS are just forming at this stage as a thickening on the medial side of each mesonephric kidney (Plate 33, Fig. 2; Plate 34, Fig. 3; Plate 35, Fig. 2). Each gonad rudiment consists of a localized thickening of visceral peritoneum (that is, a localized thickening of the splanchnic mesoderm that covers the organs of the peritoneal cavity), called the GERMINAL EPITHELIUM, and a subjacent region of condensed mesenchyme (Plate 33, Fig. 3). The gonad rudiments contain PRIMORDIAL GERM CELLS (not readily identifiable) which in mammals originate from the ENDODERM of the caudal portion of the YOLK SAC. These cells then undergo an extensive migration through the mesoderm of the yolk sac and the dorsal gut mesentery to reach the gonad rudiments.

Return to the level where the mesonephric ducts connect to the urogenital sinus (Plate 36, Fig. 1) and trace sections posteriorly. Shortly after the urogenital sinus fades out, note that a small duct emerges from the dorsal (lower) side of each mesonephric duct (Plate 36, Fig. 3). These ducts are the URETERS. Continue tracing sections posteriorly, noting that the ureters separate from the mesonephric ducts and that each eventually expands as the RENAL PELVIS (Plate 37, Fig. 1). The distinct epithelial walls of these structures (that is, the layer next to the lumen) later bud repeatedly forming the MAJOR and MINOR CALYCES and COLLECTING TUBULES of the METANEPHRIC KIDNEYS. The very dark mass of cells just outside each renal pelvis consists of condensed NEPHROTOME (Plate 37, Fig. 1) which forms the SECRETORY TUBULES of the METANEPHRIC KIDNEYS. Thus development of the metanephric kidney involves both epithelial and mesenchymal components. It has been demonstrated experimentally that neither of these components can develop in absence of the other. Thus development of the metanephric kidneys depends on EPITHELIAL-MESENCHYMAL INTERACTIONS (INDUCTIONS) which are likewise involved in development of many other structures whose rudiments consist of epithelial and mesenchymal components.

4. Circulatory system. Aids: Figs. N, O, P; A, Figs. 606, 607; M, Figs. 149, 150; PA, Figs. 66-68.

a. Arterial system. Begin with the most posterior section through the BULBUS CORDIS (Plate 31, Fig. 1) and trace sections *anteriorly.* Observe that the cavity of the bulbus cordis becomes separated into two cavities by the approximation of two ingrowths of the bulbus cordis wall. These ingrowths are the BULBAR RIDGES (Plate 30, Fig. 2); in more *anterior* sections they are fused as the BULBAR SEPTUM (Plate 30, Fig. 1) which separates the bulbus cordis into two vessels. The vessel to the *right* (apparent left) is the AORTIC TRUNK (ASCENDING AORTA); the vessel to the *left* is the PULMONARY TRUNK. At about this level note that the pulmonary trunk extends dorsad so as to be cut frontally, and is continuous with the two SIXTH AORTIC ARCHES (Plate 30, Fig. 1). Usually the *left* (apparent right) sixth aortic arch is larger in

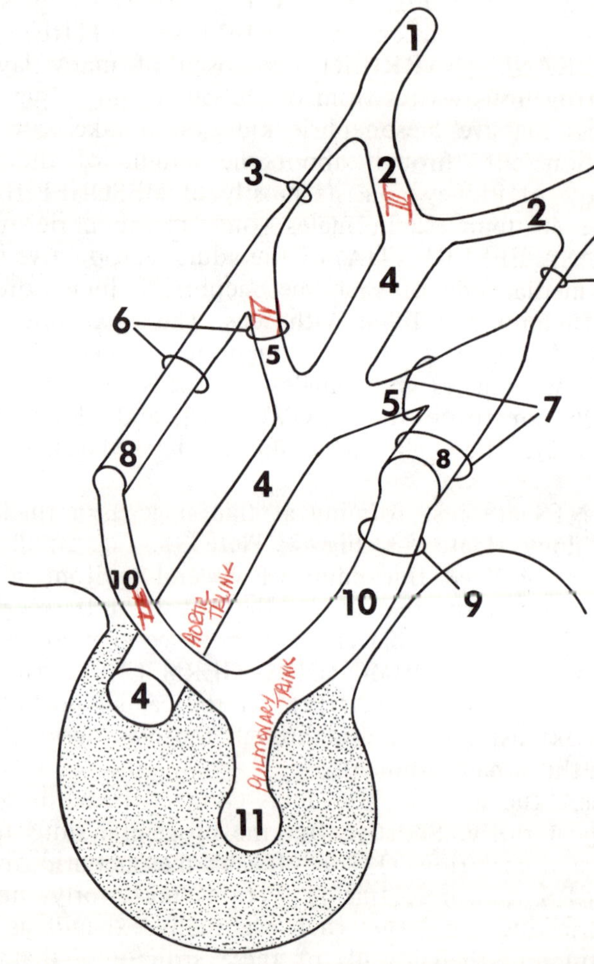

Fig. N. Reconstruction of the aortic arches of the 10-mm pig embryo. (Because this reconstruction is made directly from transverse sections, the *right* side of the embryo is drawn on the apparent left side of the reconstruction and the *left* side of the embryo is drawn on the apparent right side of the reconstruction.)

1. Internal carotid artery
2. Third aortic arch
3. Ductus caroticus
4. Aortic trunk
5. Fourth aortic arch
6. Two components of adult right subclavian artery
7. Two components of adult arch of the aorta
8. Dorsal aorta
9. Ductus arteriosus
10. Sixth aortic arch
11. Pulmonary trunk

diameter than the *right.* Continue tracing sections *anteriorly* and observe that the sixth aortic arches join the DORSAL AORTAE (Plate 29, Figs. 2, 1). The distal end of the *left* sixth aortic arch (that is, the portion which connects to the *left* dorsal aorta) is the DUCTUS ARTERIOSUS (Fig. N); it persists until birth as an important blood channel which functions as a shunt, enabling much of the blood to bypass the functionless lungs. After birth the ductus arteriosus is converted into a fibrous ligament, the LIGAMENTUM ARTERIOSUM. The distal end of the *right* sixth aortic arch soon degenerates. At the level where the sixth aortic arches join the dorsal aortae, the *right* (apparent left) dorsal aorta is usually smaller than the *left.* This level of the *right* dorsal aorta becomes part of the adult RIGHT SUBCLAVIAN ARTERY, carrying blood primarily to the right foreleg, whereas this level of the *left* dorsal aorta becomes part of the adult ARCH OF THE AORTA, carrying blood to most regions of the body (Fig. N).

Return to the level where the pulmonary trunk gives rise to the sixth aortic arches (Plate 30, Fig. 1) and trace sections *posteriorly.* As the sixth aortic arches fade out, identify the PULMONARY ARTERIES (Plate 30, Fig. 2). These arteries originated as caudal outgrowths from the sixth aortic arches. Try to trace the pulmonary arteries back towards the lung buds.

Return to the level of the bulbar septum (Plate 30, Fig. 1) and trace sections *anteriorly,* noting the disappearance of the pulmonary trunk after it gives rise to the sixth aortic arches (Plate 29, Fig. 3). (The FIFTH AORTIC ARCHES are rudimentary and have degenerated by this stage.) Identify the FOURTH AORTIC ARCHES (Plate 28, Fig. 3), continuous with the aortic trunk, and note that they quickly join the dorsal aortae (Plate 28, Figs. 3, 2). Typically the *left* (apparent right) fourth aortic arch is larger than the *right.* The former is incorporated into the adult ARCH OF THE AORTA; the latter into the adult RIGHT SUBCLAVIAN ARTERY (Fig. N). At about this level observe that the aortic trunk terminates cranially by dividing into the two THIRD AORTIC ARCHES (Plate 28, Fig. 2). (The FIRST and SECOND AORTIC ARCHES have degenerated or are degenerating by this stage.) At about this level try to identify a pair of small vessels extending ventrad from the third aortic arches. These vessels are the EXTERNAL CAROTID ARTERIES; attempt to trace them into the developing lower jaw (that is, mandibular processes).

Return to the level where the third aortic arches arise from the aortic trunk (Plate 28, Fig. 2) and trace sections *anteriorly.* Note that the third aortic arches join the dorsal aortae (Plate 27, Fig. 1). The third aortic arches form the adult COMMON CAROTID ARTERIES and part of the adult INTERNAL CAROTID ARTERIES. The remainder of the internal carotid arteries is formed from an extension of each dorsal aorta cranial to the connection between the third aortic arch and dorsal aorta (Fig. N). Observe that the portion of each dorsal aorta lying between its connection to the fourth and third aortic arches is considerably reduced in diameter (Fig. N; Plate 27, Figs. 1-3). This level of each dorsal aorta, the DUCTUS CAROTICUS, will soon completely degenerate. Following this degeneration, blood passing from the aortic trunk into the *third* aortic arches will flow cranially through the INTERNAL CAROTID ARTERIES, whereas blood passing from the aortic trunk into the *fourth* aortic arches will flow caudally through the RIGHT SUBCLAVIAN ARTERY and the ARCH OF THE AORTA.

Continue tracing sections *anteriorly* following the internal carotid arteries (that is, the cranial extensions of the dorsal aortae). They are quickly cut frontally (Fig. O, level a; Plate 26, Fig. 3) and disappear a few sections more *anteriorly.* Return to the level where the internal carotid arteries are cut frontally (Fig. O, level a; Plate 26, Fig. 3). Shift your attention to the lower end of each internal carotid artery (that is, the end nearest the diencephalon) and trace sections *posteriorly.* The internal carotid arteries gradually move downwards towards the diencephalon (Fig. O, level b; Plate 27, Figs. 1-3) where each joins another vessel cut frontally. These latter vessels are the cranial ends of the internal carotid arteries (Fig. O, level c). Each is continuous at its lower end with a small ANTERIOR CEREBRAL ARTERY (Plate 27, Fig. 3). Continue tracing sections *posteriorly* watching the anterior cerebral arteries. They are difficult to follow since they are just developing, but they can usually be identified in more *posterior* sections lateral to the diencephalon (Plate 28, Fig. 3).

Return to the level where the internal carotid and anterior cerebral arteries on each side are continuous (Plate 27, Fig. 3) and trace sections *anteriorly* watching the cranial end of each

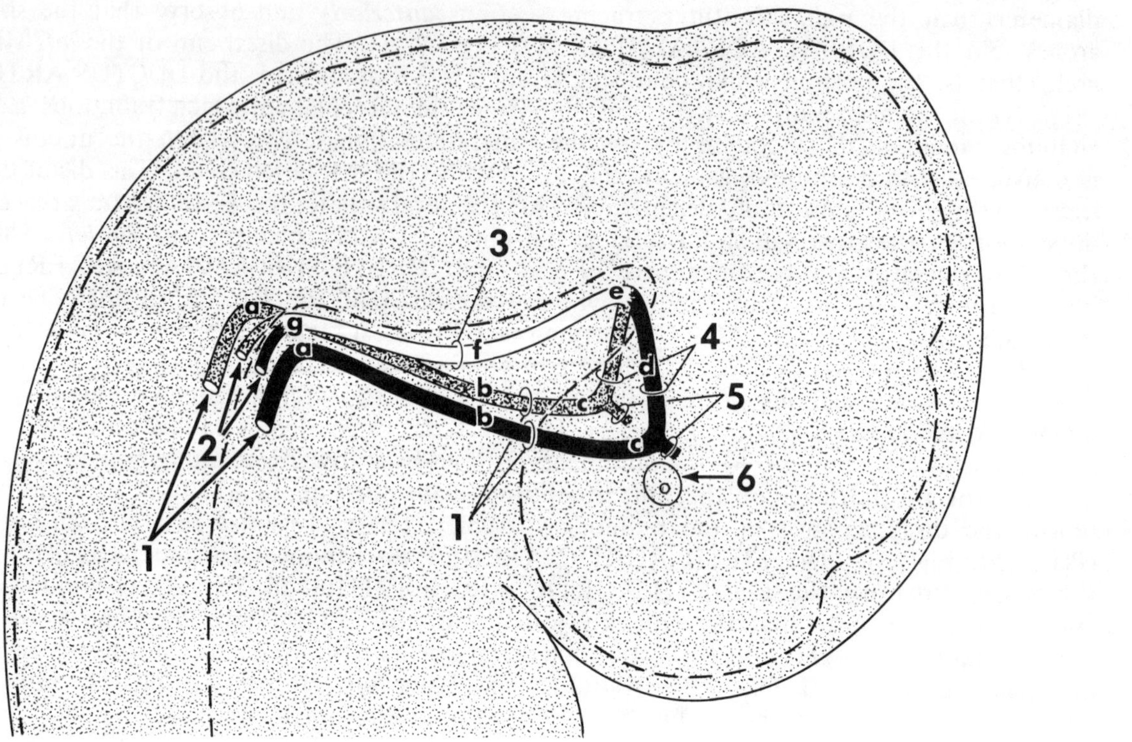

Fig. O. Schematic drawing of the arteries in the head of the 10-mm pig embryo. Arteries on the right side of the embryo are indicated by heavy black lines and on the left side by stippling. The basilar artery (3) is in the midline. Letters a-g indicate specific levels described in text. The dashed line indicates the boundaries of the brain and spinal cord; the outer line the boundaries of the head.

1. Internal carotid arteries
2. Vertebral arteries
3. Basilar artery
4. Posterior communicating arteries
5. Anterior cerebral arteries
6. Eye

internal carotid artery (that is, the portion which is cut frontally). The POSTERIOR COMMUNICATING ARTERIES quickly appear in place of these latter frontally cut vessels (Fig. O, level d; Plate 27, Fig. 1). (The anterior cerebral and posterior communicating arteries will later become components of the ARTERIAL CIRCLE OF WILLIS, a complex of blood vessels delivering blood to the ventral surface of the brain.) Continue tracing sections *anteriorly* following the posterior communicating arteries. They first lie lateral to the infundibulum (Plate 26, Fig. 3) and then ventral to the mesencephalon (Plate 26, Figs. 2, 1; Plate 25, Fig. 3). Note that they approach a vessel lying in the midline just beneath the rhombencephalon, the BASILAR ARTERY, and the three vessels become continuous (Fig. O, level e). (The exact region where the two posterior communicating arteries and basilar artery become continuous is difficult to identify.)

Reverse direction and trace sections *posteriorly* following the basilar artery. As the rhombencephalon begins to fade out, identify a single vessel lying in the midline just above the rhombencephalon (Plate 26, Figs. 1, 2). This vessel is part of the basilar artery; in more *posterior* sections the rhombencephalon fades out and the two parts of the basilar artery join and are cut frontally (Fig. O, level f; Plate 27, Fig. 1). In more *posterior* sections the basilar artery fades out.

Return to the level where the basilar artery is cut frontally (Fig. O, level f; Plate 27, Fig. 1) and trace sections *anteriorly* following the upper portion of the basilar artery. Note that it lengthens dorso-ventrally and is continuous with two vessels, the VERTEBRAL ARTERIES (Fig. O, level g; Plate 26, Fig. 1), which extend dorsad. The vertebral arteries are irregularly shaped in this region. Trace sections *posteriorly* following the vertebral arteries beneath the spinal cord. (In the adult each vertebral artery receives its blood directly from a SUBCLAVIAN ARTERY.)

Return to the level of the stomach (Plate 32, Fig. 3) and identify the DESCENDING AORTA. Trace sections posteriorly. At about the level that the stomach fades out, or, more commonly, in slightly more posterior sections, identify a single midline vessel continuous with the descending aorta ventrally, the CELIAC ARTERY (see A, Fig. 622). It originated by fusion of a pair of VITELLINE ARTERIES at an earlier stage. Continue tracing sections posteriorly and identify paired vessels continuous with the descending aorta dorso-laterally; these are the INTERSEGMENTAL ARTERIES. (The adult LEFT SUBCLAVIAN ARTERY is derived exclusively from a left intersegmental artery at the level of the foreleg buds. The adult RIGHT SUBCLAVIAN ARTERY is derived from a right intersegmental artery at the level of the foreleg buds [Plate 31, Fig. 3], part of the right dorsal aorta, and part of the right fourth aortic arch; see Fig. N.)

Continue to trace sections posteriorly and identify small ventro-lateral branches of the descending aorta. These are the MESONEPHRIC (LATERAL) ARTERIES, which carry blood to the glomeruli of the mesonephric kidneys (see A, Fig. 623). In sections at about the level of the caudal end of the liver, identify a large midline vessel continuous with the descending aorta ventrally (Plate 34, Fig. 3). This vessel is the SUPERIOR MESENTERIC ARTERY (see A, Fig. 606); it also originated by fusion of a pair of VITELLINE ARTERIES. The distal end of the superior mesenteric artery can be identified within the umbilical cord at about this level (Plate 34, Fig. 3). Trace sections *posteriorly* until the two portions of the superior mesenteric artery join and are cut frontally (Plate 35, Fig. 2).

Quickly trace sections posteriorly until the descending aorta is cut frontally due to the tail flexure (Plate 37, Fig. 2). In more posterior sections the descending aorta disappears. Return to the level at which the descending aorta is cut frontally (Plate 37, Fig. 2), and trace sections *anteriorly* following the lower part of this vessel. Identify the right and left UMBILICAL ARTERIES continuous with the descending aorta (Plate 37, Fig. 2). A smaller midline portion of the descending aorta can be traced into the tail as the CAUDAL ARTERY (Plate 37, Figs. 2, 1). Try to identify a small lateral branch from each umbilical artery; these branches are the EXTERNAL ILIAC ARTERIES (Plate 37, Fig. 1). Continue tracing sections *anteriorly.* The umbilical arteries first lie at the base of the hindleg buds (Plate 36, Figs. 3, 2). This portion of the umbilical arteries forms part of the adult INTERNAL ILIAC ARTERIES. In more *anterior* sections the umbilical arteries can be identified within the umbilical cord (Plate 35, Figs. 2, 1; Plate 34, Figs. 3, 1).

b. Heart. Return to the most posterior section through the bulbar septum (Plate 30, Fig. 1) and trace sections posteriorly. Identify the RIGHT and LEFT ATRIA, and the muscular LEFT VENTRICLE (Plate 30, Figs. 2, 3). Continue tracing sections posteriorly and watch the two atria approach each other and become continuous (Plate 31, Fig. 1). At about this level the bulbus cordis and muscular RIGHT VENTRICLE are continuous. The cavities of the right and left atria are partially separated at this level by a partition, the SEPTUM PRIMUM (SEPTUM I). This septum is incomplete dorsally as the FORAMEN SECUNDUM (FORAMEN II). (Earlier there existed a more ventral opening in the septum primum, the FORAMEN PRIMUM [FORAMEN I], which is now closed.) In slightly more posterior sections the septum primum appears to completely separate the two atria (Plate 31, Fig. 2). Continue tracing sections posteriorly and observe that the cavities of the atrium and ventricle on each side are continuous via an ATRIO-VENTRICULAR CANAL (Plate 31, Fig. 3). (The TRICUSPID VALVE *later* develops within the *right* atrio-ventricular canal; the BICUSPID [MITRAL] VALVE develops within the *left*.) The atrio-ventricular canals lie on either side of a midline mass of lightly stained tissue, the ENDOCARDIAL CUSHION. Identify the prominent INTERVENTRICULAR SEPTUM, partially separating the cavity of the ventricle into right and left sides. At this stage the interventricular septum is incomplete dorsally as the INTERVENTRICULAR FORAMEN. (This foramen will later be closed by the SEPTUM MEMBRANACEUM derived from a downgrowth from the endocardial cushion and from part of the bulbar septum.)

Return to the level of the foramen secundum (Plate 31, Fig. 1) and trace sections posteriorly. Identify the SINUS VENOSUS lying above the *right* atrium (Plate 31, Figs. 2, 3), and the SINO-ATRIAL VALVE projecting into the cavity of the *right* atrium (Plate 31, Fig. 2). The sinus venosus is eventually absorbed by the wall of the *right* atrium as the PACEMAKER of the adult heart. The sino-atrial valve consists of right and left VALVE FLAPS which are usually in contact with one another. The *left* (apparent right) valve flap forms part of the SEPTUM SECUNDUM (SEPTUM II) which will therefore lie to the *right* (apparent left) of the septum primum. An opening, the FORAMEN OVALE, persists between the septum secundum and endocardial cushion (Plate 31, Fig. 2). The foramen ovale is staggered with respect to the opening in the septum primum (that is, the foramen secundum). The positioning of these two foramina and atrial septa is such that the septum primum acts as a flutter valve allowing blood to pass only *from the right to the left* atrium.

*c. **Venous system.*** Return again to the level of the foramen secundum (Plate 31, Fig. 1) and trace sections posteriorly. Shortly before the *left* atrium fades out, identify a single PULMONARY VEIN continuous with it (Plate 31, Fig. 3). (This vein will later branch to form four PULMONARY VEINS which open directly into the *left* atrium in the adult.)

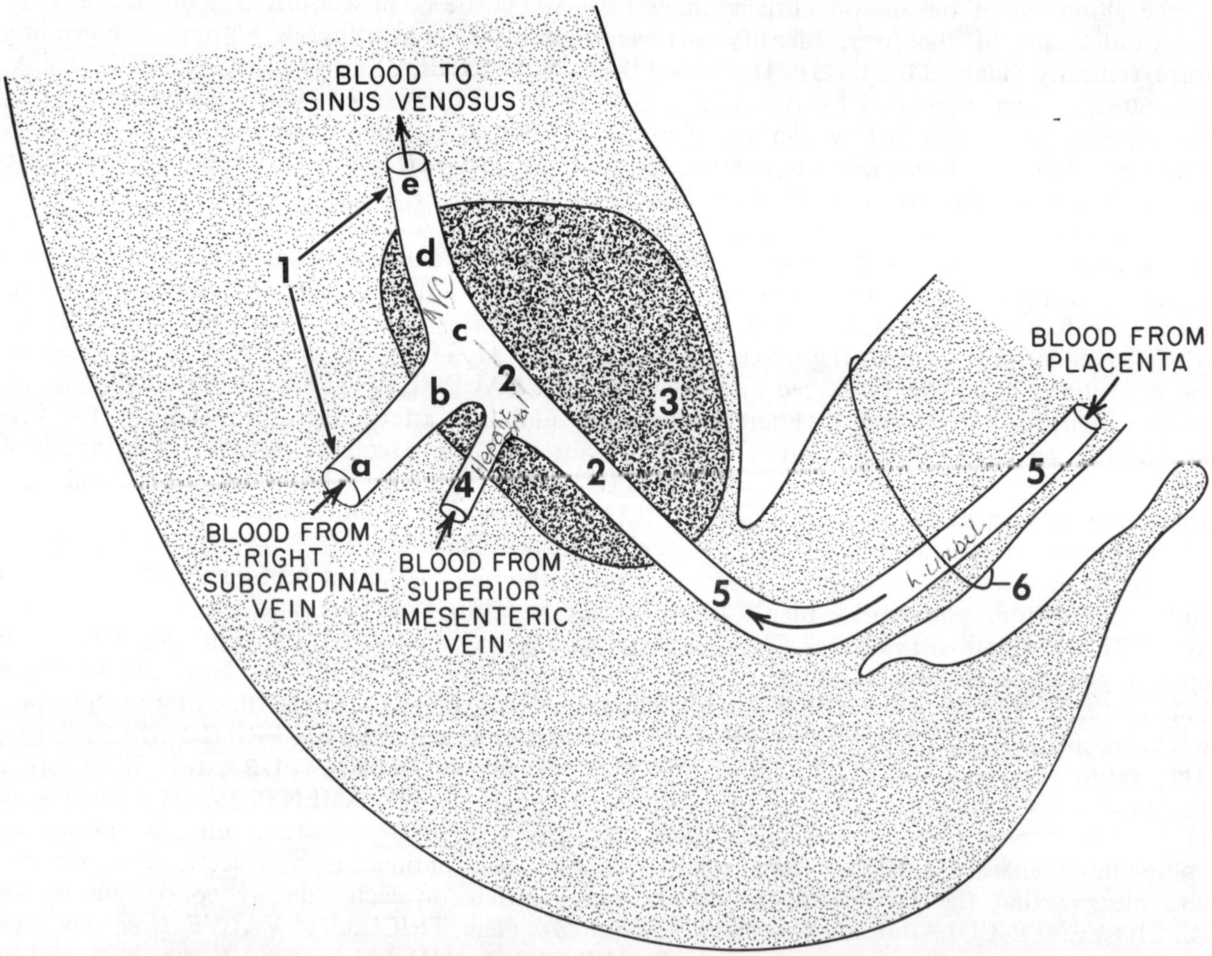

Fig. P. Schematic drawing of the major veins associated with the liver in the 10-mm pig embryo. The *right* umbilical vein is not shown. Letters a-e indicate specific levels of the inferior vena cava (1) described in text. The outer line indicates the boundaries of the caudal two-thirds of the embryo and of the umbilical cord. Arrows indicate direction of blood blow.

1. Inferior vena cava
2. Ductus venosus
3. Liver
4. Hepatic portal vein
5. Left umbilical vein
6. Umbilical cord

Quickly trace sections posteriorly until the umbilical cord can be identified (Plate 34, Fig. 1). Within the umbilical cord identify the right and left UMBILICAL (ALLANTOIC) VEINS (Plate 34, Figs. 1, 3; Plate 35, Figs. 1, 2). The *right* (apparent left) umbilical vein is smaller than the *left* since it is degenerating. After this degeneration occurs, all oxygenated blood will flow from the PLACENTA to the embryo via the *left* umbilical vein. Each umbilical vein is cut across twice. As sections are traced posteriorly the two portions of each vein become continuous and are cut frontally (Plate 35, Fig. 3; Plate 36, Fig. 1). In more posterior sections the umbilical veins fade out.

Return to the level where the umbilical veins are cut frontally (Plate 36, Fig. 1; Plate 35, Fig. 3) and trace sections *anteriorly* following the upper portion of the umbilical veins. In addition, observe the region between the two mesonephric kidneys just beneath the descending aorta. Right and left SUBCARDINAL VEINS may be located here (Plate 34, Figs. 3, 1), or the subcardinal veins may have fused to form a SUBCARDINAL ANASTOMOSIS (Plate 35). (In some embryos the subcardinal veins are fused at several localized levels forming multiple subcardinal anastomoses.) Continue tracing sections *anteriorly* and note that the *left* (apparent right) umbilical vein enters the liver (Plate 34, Fig. 1). (Usually the right umbilical vein joins the left shortly before the latter enters the liver, but in younger embryos the right umbilical vein also enters the liver.) Also note that as sections are traced *anteriorly* the *left* (apparent right) subcardinal vein fades out, and the *right* continues as part of the INFERIOR VENA CAVA (Fig. P, level a; Plate 33, Fig. 3).

Return to the level where the left umbilical vein enters the liver (Plate 34, Fig. 1). At about this level identify a vessel lying above and to the *left* (apparent right) of the duodenum. In slightly more *anterior* sections this vessel shifts to the *right* of the duodenum (Plate 33, Fig. 3). This vessel constitutes the SUPERIOR MESENTERIC VEIN, when it lies to the *left* of the duodenum (Plate 34, Fig. 1), and the HEPATIC PORTAL VEIN, when it lies in the midline and to the *right* of the duodenum (Plate 33, Fig. 3). Continue to trace sections *anteriorly* and note that the hepatic portal vein enters the right dorsal lobe of the liver, and that the inferior vena cava begins to shift ventrad towards the liver (Plate 33, Figs. 2, 1). In more *anterior* sections the inferior vena cava enters the right dorsal lobe of the liver (Fig. P, level b; Plate 32, Fig. 3). At about this level the DUCTUS VENOSUS can be identified. It is formed by the enlargement of hepatic sinusoids.

Return again to the level where the left umbilical vein enters the liver (Plate 34, Fig. 1) and trace sections *anteriorly*. Try to trace the left umbilical vein (and possibly also the right) until it joins the ductus venosus. (The exact boundary between the left umbilical vein and ductus venosus is impossible to distinguish. The latter is irregularly shaped; the former is circular in transverse section.) At about this level try to trace the hepatic portal vein until it joins the ductus venosus (this happens at a section level between Plate 33, Fig. 1 and Plate 32, Fig. 3). In slightly more *anterior* sections the inferior vena cava and ductus venosus become continuous (Fig. P, level c; Plate 32, Fig. 2). In more *anterior* sections this fused vessel narrows as the inferior vena cava (Fig. P, level d; Plate 32, Fig. 1). Continue tracing sections *anteriorly* and note that the inferior vena cava exits from the liver (Fig. P, level e) and opens into the sinus venosus.

Return to a level where the hepatic portal vein can be identified within the right dorsal lobe of the liver (Plate 33, Figs. 1, 2) and trace sections *posteriorly*. It quickly leaves the liver and lies between the dorsal and ventral pancreatic rudiments to the *right* (apparent left) of the duodenum (Plate 33, Fig. 3). Note that it shifts to the left of the duodenum to become the superior mesenteric vein (Plate 34, Fig. 1), one of the main branches of the hepatic portal vein in the adult. Trace sections posteriorly following the superior mesenteric vein. It soon joins a prominent COMMON VITELLINE VEIN (Plate 34, Fig. 3). (The superior mesenteric, hepatic portal, and common vitelline veins are derived from a pair of VITELLINE VEINS present earlier. The superior mesenteric vein is derived from the proximal part of the *left* vitelline vein, the hepatic portal vein from the proximal part of the *right* vitelline vein and from a cross connection between right and left vitelline veins, and the common vitelline vein from an anastomosis between the distal ends of the right and left vitelline veins.) In more posterior sections the superior mesenteric and common vitelline veins can be traced into the umbilical cord and are cut frontally—first the common vitelline vein, and then the superior mesenteric vein (Plate 35, Fig. 2). These veins then disappear. Return to the levels at which the superior mesenteric (Plate 35, Fig. 2) and common vitelline veins are cut frontally and trace sections *anteriorly* following the lower portion of each vein within the umbilical cord. Trace these vessels as far as possible.

Return to a section where the PRECARDINAL VEINS can be first identified (Plate 25, Fig. 3). At about this level these veins are cut frontally. Trace sections *posteriorly* following the lower portion of each precardinal vein until it fades out ventro-lateral to the diencephalon. (The adult SUPERIOR VENA CAVA is derived in part from a portion of the *right* precardinal vein. The adult INTERNAL JUGULAR VEINS are derived from the cranial ends of both precardinal veins.)

Return again to a section where the precardinal veins can be first identified (Plate 25, Fig. 3) and quickly trace sections posteriorly following the *upper* portion of each precardinal vein. Note that the precardinal veins are closely approximated to the nodosum ganglia (Plate 27, Fig. 3). In more posterior sections the precardinal veins gradually move ventrad (Plate 28, Fig. 3; Plate 29, Fig. 3; Plate 30, Fig. 3). At about the level of the foreleg buds these vessels elongate dorso-ventrally (Plate 31, Figs. 1, 2), indicating that sections are now cutting through the COMMON CARDINAL VEINS. (The adult SUPERIOR VENA CAVA is derived from the *right* common cardinal vein as well as from the *right* precardinal vein. The left common cardinal vein forms the adult CORONARY SINUS and OBLIQUE VEIN OF THE LEFT ATRIUM.) The *right* common cardinal vein quickly enters the sinus venosus (Plate 31, Fig. 2). The *left* common cardinal vein likewise enters the sinus venosus, but in more posterior sections, by means of a narrow connection beneath the trachea (use high magnification and carefully trace sections to observe this connection). Try to identify the irregularly shaped SUBCLAVIAN VEINS just developing at the base of the foreleg buds (Plate 30, Fig. 3; Plate 31, Fig. 1).

Return to the level where the *right* common cardinal vein enters the sinus venosus (Plate 31, Fig. 2) and trace sections posteriorly. Identify the POSTCARDINAL VEINS quickly appearing in place of the common cardinal veins (Plate 31, Fig. 3). Continue tracing sections posteriorly following the postcardinal veins as far as possible. Note that they first lie dorso-lateral to the mesonephric kidneys (Plate 32, Fig. 1). In more posterior sections they are cut across frontally (Plate 37, Fig. 2). Reverse direction and trace sections *anteriorly* following the lower portion of each postcardinal vein located at the base of a hindleg bud (Plate 37, Fig. 2) into the tail (Plate 36, Fig. 3). (The *left* postcardinal vein later primarily degenerates. The proximal part of the *right* postcardinal vein forms part of the adult AZYGOS VEIN.)

C. SUMMARY OF GERM LAYER DERIVATIVES FORMED SINCE THE 72-HOUR CHICK EMBRYO STAGE

Ectoderm

autonomic ganglia
brachial plexus
cranial ganglia:
 VII. geniculate
 VIII. acoustic
 IX. petrosal
 X. nodosum
cranial nerves:
 VI. abducens
 VIII. auditory
 X. vagus
 XI. spinal accessory
 XII. hypoglossal
Froriep's ganglia
glioblasts
lamina terminalis
pigment granules (within pigmented retinas)
rudiments of the semicircular canals
spinal nerves

Mesoderm

choroid
cloacal septum
eye muscle rudiments
gonad rudiments
heart partitions:
 bulbar ridges and septum
 endocardial cushion
 interventricular septum
 septum primum
 septum secundum
mesenteries:
 falciform ligament
 mesocolon
mesonephric kidneys
pericardium
peritoneum
pleura
renal pelvis
sclera
septum transversum
sino-atrial valve
ureters
Wharton's jelly

Endoderm

caudal limb of the intestinal loop
colon
common bile duct
cranial limb of the intestinal loop
cystic duct
eparterial bronchus
gall bladder
hepatic cords
hepatic duct
larynx
pancreatic ducts and rudiments
rectum
thyroid gland
trachea
urogenital sinus

Ectoderm and Mesoderm

choroid plexus
genital eminence
mammary ridges
nasal processes
umbilical cord

Endoderm and Mesoderm

arytenoid swellings
epiglottis
tongue rudiments

V

Early Embryology of the Frog

Amphibian eggs are relatively large and can be readily obtained. Many outstanding experimental embryologists (including Hans Spemann, Ross Harrison, and their many students) took advantage of these facts when they began to experiment on developing vertebrate embryos by removing parts, adding parts, and recombining parts by microsurgery. Since so much of classical experimental embryology involved experiments on early amphibian embryos, it is essential in order to appreciate experimental analyses that you have an understanding of the structure of these embryos and how this structure originated. The study of early amphibian development will also demonstrate the basic similarities in developmental events and processes in amphibians, birds, and mammals.

A. OOGENESIS AND FERTILIZATION

OOGENESIS (development of the ovum) begins in the paired OVARIES of the mature female frog. During the breeding season (which begins in the spring), each ovary consists of a sac containing a cluster of spherical structures called FOLLICLES (Fig. Q). Each follicle consists of a large central cell containing a lot of YOLK in its cytoplasm, the PRIMARY OOCYTE, surrounded by a layer of much smaller, flattened cells called FOLLICLE CELLS. The primary oocyte contains a large nucleus, the GERMINAL VESICLE. The VITELLINE MEMBRANE lies between the follicle cells and the primary oocyte. A thin sheath of connective tissue, the THECA FOLLICULI EXTERNA, forms the surface layer of the ovary. Another sheath, the THECA FOLLICULI INTERNA, partially surrounds each follicle, but is lacking in the region where the follicle contacts the theca folliculi externa; at this region OVULATION (the rupture of the follicle and release of its contained oocyte) occurs.

Each ovary also contains cells called OOGONIA. These cells undergo rapid mitotic divisions, increasing in number. After the breeding season is completed (that is, in the autumn), several hundred oogonia within each ovary lose the ability to divide mitotically and enlarge slightly as PRIMARY OOCYTES. Primary oocytes enter the PROPHASE stage of the FIRST MEIOTIC DIVISION, but remain in this stage until the following spring. Each newly formed primary oocyte soon becomes surrounded by a single layer of follicle cells, forming an ovarian follicle, and slowly enlarges due to the accumulation of yolk.

When spring arrives, the fully grown oocytes undergo ovulation in response to hormones secreted by the ANTERIOR PITUITARY GLAND (ADENOHYPOPHYSIS). During ovulation each oocyte is slowly squeezed through the follicular wall, at the region where the theca folliculi interna is lacking (Fig. Q), and enters the body cavity (coelom) of the female. Each spring approximately 2,000-20,000 oocytes ovulate in a single female, depending on the species. During ovulation each primary oocyte completes the first meiotic division, forming a FIRST POLAR BODY and SECONDARY OOCYTE. Both of these structures are contained within the vitelline membrane formed earlier while the primary oocyte was in the ovary. Cilia on the lining of the coelom beat towards the OSTIUM of the OVIDUCT and propel the secondary

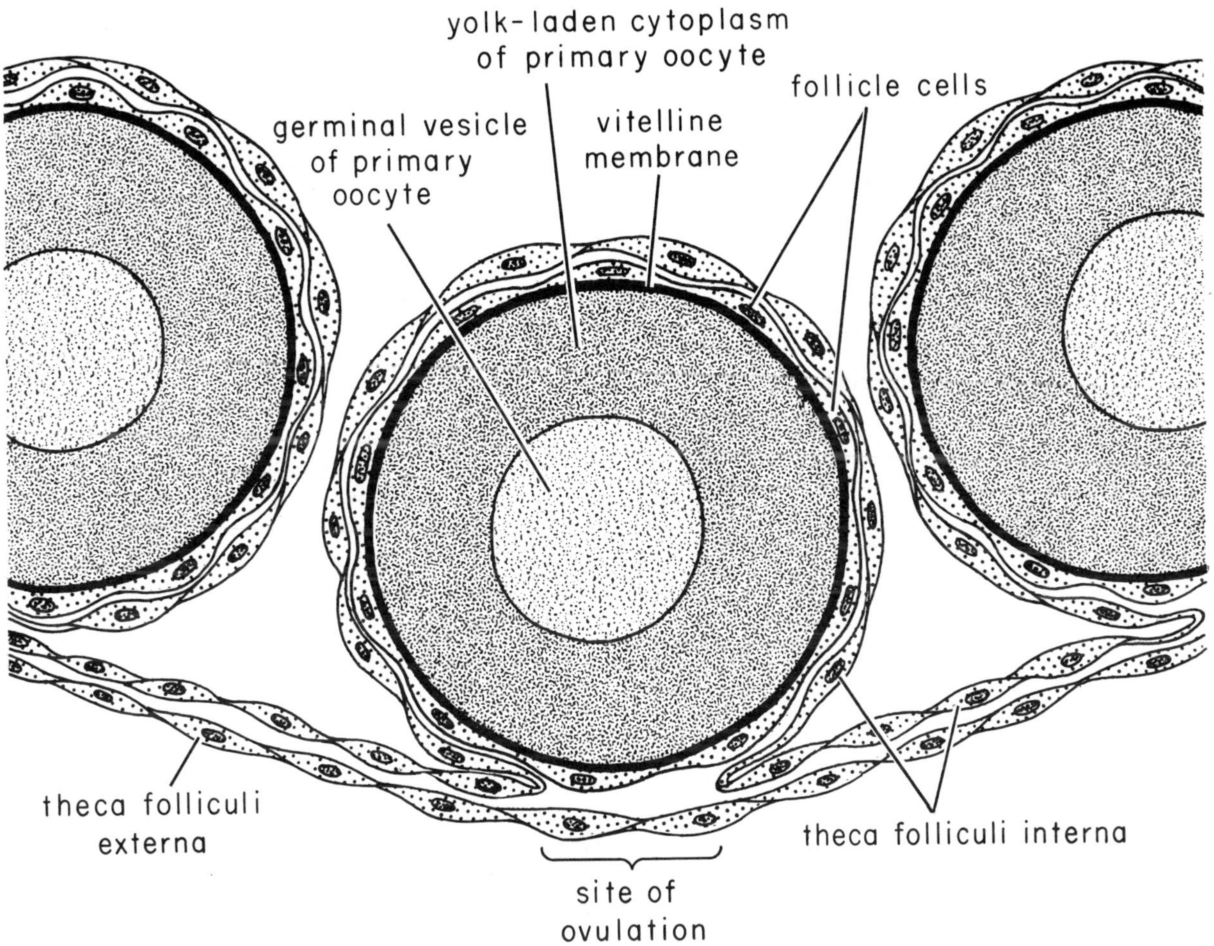

Fig. Q. Schematic drawing of a section through three ovarian follicles of a mature female frog.

oocytes into this opening. The SECOND MEIOTIC DIVISION is initiated by each secondary oocyte at about the time that it enters the oviduct, but this division is arrested in the METAPHASE stage. As the secondary oocytes pass through the oviduct, a multilayered, gelatinous EGG CAPSULE is secreted outside of the vitelline membrane by the cells lining the oviduct.

FERTILIZATION occurs externally as the secondary oocytes are spawned (shed) by the female into the water. As a successful sperm contacts and penetrates each secondary oocyte, the second meiotic division is completed, resulting in formation of a SECOND POLAR BODY and a MATURE OVUM containing the FEMALE PRONUCLEUS. The nucleus of the successful sperm within the ovum enlarges as the MALE PRONUCLEUS and the male and female pronuclei unite to complete the process of fertilization.

B. FORMATION OF THE GRAY CRESCENT

Aids: Fig. R; H, Fig. 46.

At the time of ovulation the outer portion of the egg (the CORTEX) contains a distinct pattern of pigmentation (Fig. Ra). About two-thirds of the cortex is heavily pigmented; the remaining portion contains almost no pigment. The uppermost part of the pigmented portion is the ANIMAL POLE. This pole corresponds to the *cranial* end of the future embryo. The half of the egg which contains the animal pole is the ANIMAL HEMISPHERE. The VEGETAL POLE lies directly opposite the animal pole. This pole corresponds to the *caudal* end of the future embryo. The half of the egg which contains the vegetal pole is the VEGETAL HEMISPHERE.

Following contact by and entrance of the successful sperm, the pigmented cortex shifts towards the site of sperm entrance and away from the side of the egg opposite the sperm entrance point (Fig. Rb). This reduces the pigmentation of a crescent-shaped area opposite the point of sperm entrance. This crescent-shaped area between the heavily pigmented cortex above and the essentially nonpigmented cortex below constitutes the GRAY CRESCENT (Figs. Rb, Rc). A plane passing through the animal and vegetal poles and through the center of the gray crescent corresponds to the midsagittal plane of the future embryo (Fig. Rc).

C. CLEAVAGE AND BLASTULATION

Aids: Fig. R; Plate 38, Fig. 1; H, Figs. 46, 47; M, Figs. 41-43.

CLEAVAGE consists of a series of rapid mitotic divisions which result in BLASTULATION (that is, the formation of a BLASTULA consisting of a group of cells called BLASTOMERES surrounding a cavity, the BLASTOCOEL). CLEAVAGE FURROWS pass through the entire egg so cleavage is classified as TOTAL (HOLOBLASTIC). However, furrows pass through the vegetal hemisphere much more slowly than through the animal hemisphere presumably because the former contains far more yolk than the latter. A rough time table of the first three cleavages for the frog, *Rana pipiens*, is as follows (Figs. Rd-Rf): (1) first cleavage: meridional (that is, passes through both animal and vegetal poles) and usually bisects the gray crescent (that is, passes through the midsagittal plane of the future embryo), 2½ hours postfertilization; (2) second cleavage: meridional, at right angles to the first, 3½ hours after fertilization; (3) third cleavage: horizontal, displaced towards the animal pole due to the yolk content of the vegetal hemisphere, producing four smaller animal hemisphere cells, the MICROMERES, and four larger vegetal hemisphere cells, the MACROMERES, 4 hours postfertilization. At the end of cleavage (approximately 16 hours postfertilization) the egg is in the BLASTULA stage (Fig. Rg). Numerous small blastomeres are present in the animal hemisphere, while the vegetal hemisphere contains a lesser number of larger blastomeres. At the end of cleavage the blastula is only slightly larger than the newly fertilized egg.

Examine models, or preferably living eggs, showing early cleavage stages. Note the position of early cleavage furrows (that is, whether meridional or horizontal), the difference in size of the micromeres and macromeres, and difference in pigmentation of animal and vegetal hemisphere cells. Try to identify the GRAY CRESCENT; the region of the gray crescent which is broadest will become the *dorsal* surface of the embryo, and the opposite side the *ventral* surface.

Examine a section of the frog blastula closely resembling Plate 38, Fig. 1. Identify the main cavity, the BLASTOCOEL. Note that it is displaced towards one side of the blastula, the ANIMAL POLE; thus it is contained within the ANIMAL HEMISPHERE. The blastocoel is filled with a fluid which coagulated during preparation of your slide. Note that the wall of the blastula is composed of distinct cells, the BLASTOMERES. Identify their NUCLEI. Spaces between blastomeres in the VEGETAL HEMISPHERE are shrinkage spaces produced during preparation of your slides.

Examine the ANIMAL HEMISPHERE. It has the following characteristics: (1) there is a heavily pigmented cortex; pigmentation is most intense at the animal pole and grades off progressively towards the vegetal pole; (2) it is four or five cells thick, with smaller cells and fewer cell layers at the animal pole and with a progressive increase in cell size and number of cell layers towards the vegetal pole; (3) the blastomeres contain very little yolk.

Examine the VEGETAL HEMISPHERE. Its characteristics are exactly the opposite: (1) a pigmented cortex, if present at all, is much less evident than in the animal hemisphere; (2) the blastomeres are very large and few in number, indicative of less frequent cleavage; (3) the blastomeres are packed with yolk.

Try to identify the area of the GRAY CRESCENT. In your sections it usually lies to either the left or right side of the blastocoel and also slightly ventral to it. It has the following characteristics (compare sides indicated by letters D and V in Plate 38, Fig. 1): (1) the pigmented cortex is thinner in the area of the gray crescent than on the opposite side; (2) the blastocoel

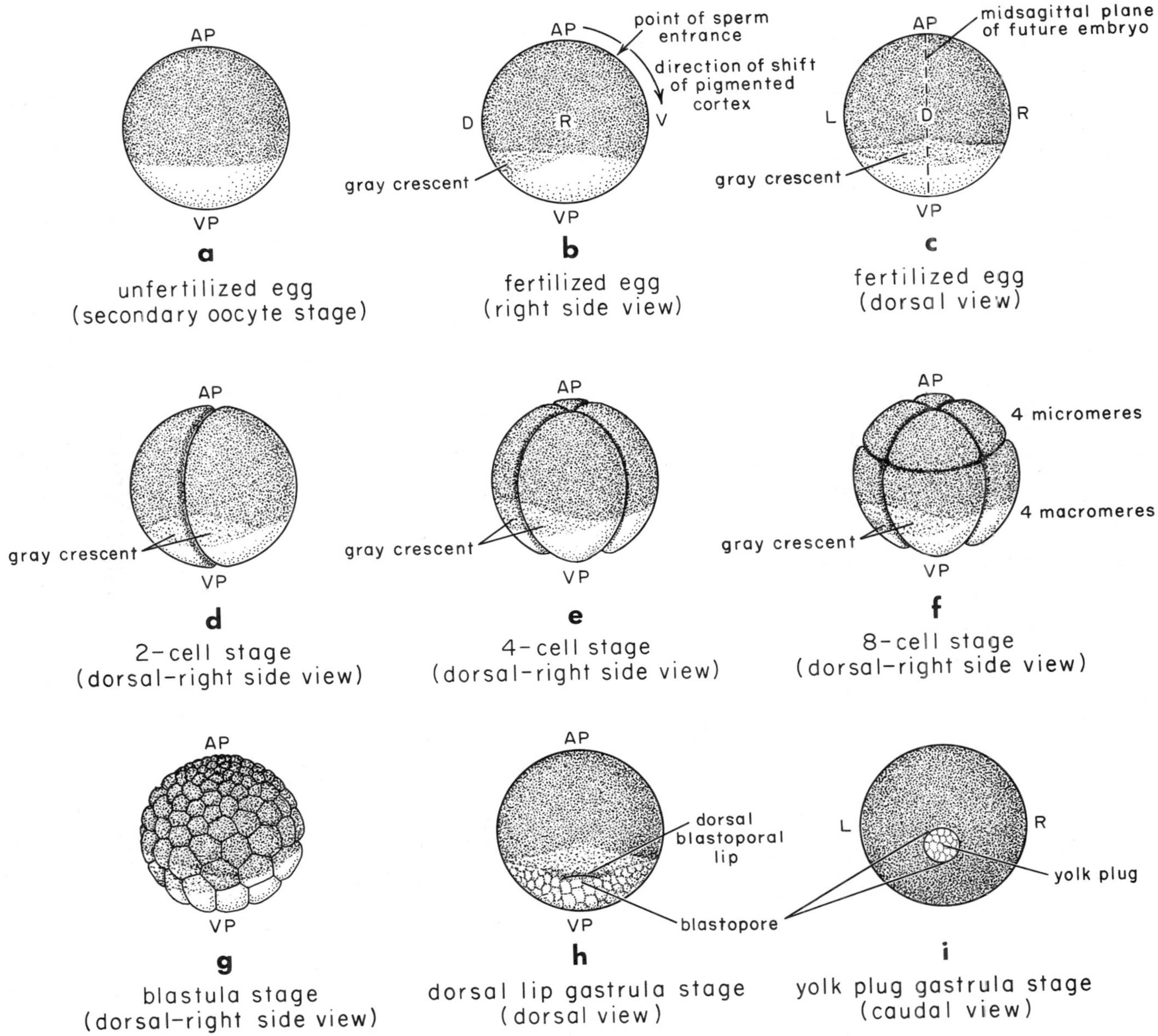

Fig. R. Drawings of early developmental stages of the frog. AP, animal pole; D, dorsal side; L, left side; R, right side; V, ventral side; VP, vegetal pole.

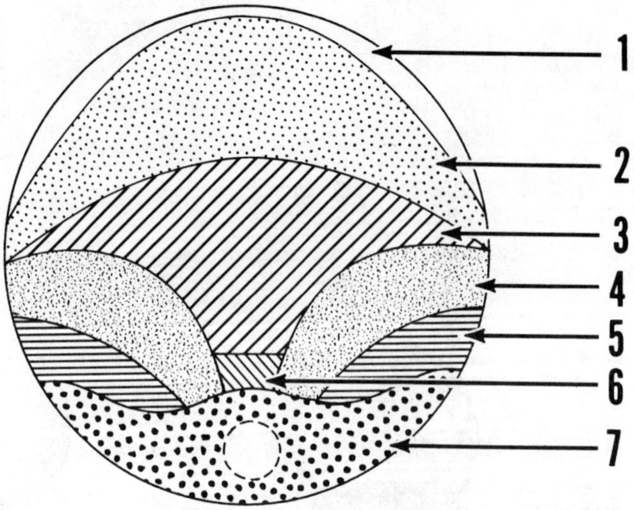

Fig. S. Prospective fate map of the frog blastula. The approximate site of blastopore formation is indicated by a circular dashed line.

1. Prospective ectoderm
2. Prospective neural plate
3. Prospective notochord
4. Prospective segmental plate mesoderm
5. Prospective lateral plate mesoderm
6. Prospective head mesenchyme
7. Prospective endoderm

lies nearer the surface on the gray crescent side than on the opposite side (that is, the wall of the blastula is thinner on the gray crescent side than on the opposite side).

The blastula consists of a mosaic of cellular areas, each of which will normally produce a certain structure during subsequent development. In other words each area of cells has a certain PROSPECTIVE FATE which will be realized in normal development. In blastulae of some chordate species (Urochordata or tunicates) the outlines of these cellular areas can be determined directly because the cytoplasm of cells within certain areas is colored differently. But in most cases it is necessary to determine the prospective fate of each cellular area indirectly by marking experiments. In amphibians VITAL DYES have been most frequently used for this purpose. Several areas of the blastula are stained with a vital dye. It is then observed what structure or structures are formed from each stained area, and with the aid of this information a PROSPECTIVE FATE MAP is constructed such as the one in Fig. S. The amphibian fate map should be carefully compared with the one for the chick (Fig. H). A prospective fate map indicates the location of specific groups of cells prior to the onset of GASTRULATION. During gastrulation these groups of cells are shifted in an orderly way into appropriate positions, which will enable them to cooperate and interact in formation of tissues and organs. The blastocoel appears to be essential in many species to provide a space into which certain groups of cells can move *en masse* or individually during gastrulation.

D. GASTRULATION

Aids: Figs. R-T; Plate 38, Figs. 2, 3; B, Fig. 125; H, Figs. 48, 49, 53; M, Figs. 44-49.

During gastrulation cells originally located on the surface of the blastula turn inward or undergo INVOLUTION to move into the interior. A depression, the BLASTOPORE, begins to form just below the gray crescent as cells initiate involution (Fig. Rh). Simultaneously, a liplike structure, the DORSAL BLASTOPORAL LIP, forms just above the blastopore. With formation of the blastopore and dorsal blastoporal lip the blastula is transformed into a GASTRULA. Cells continue to involute over the dorsal blastoporal lip with further development and involution progressively occurs laterally and ultimately ventrally as well. This results in formation of a

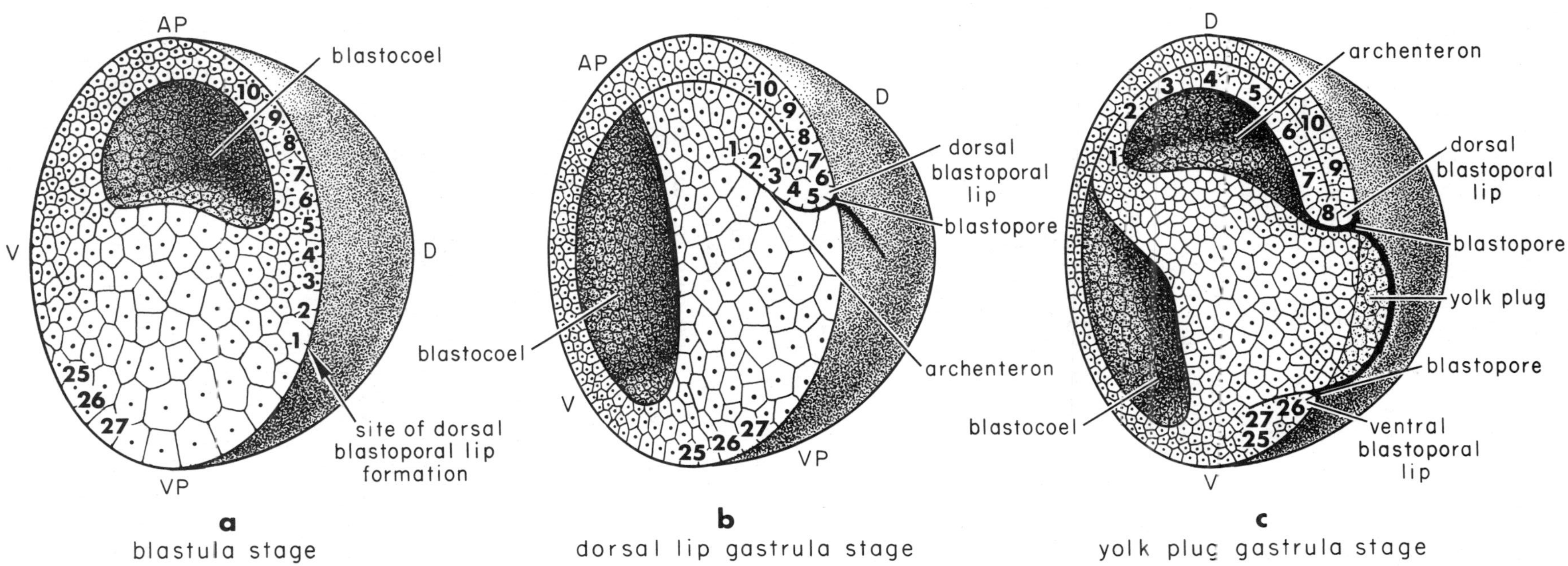

Fig. T. Schematic diagrams of the cut surface of right halves of blastula and gastrula stages of the frog. AP, animal pole; D, dorsal side; V, ventral side; VP, vegetal pole; numbers 1-10 indicate specific areas of the blastula which undergo involution over the dorsal blastoporal lip; numbers 25-27 indicate specific areas which undergo involution over the ventral blastoporal lip.

circular blastopore containing a mass of yolk-filled endodermal cells called the YOLK PLUG (Fig. Ri). The circular blastopore is surrounded by DORSAL, LATERAL, and VENTRAL BLASTOPORAL LIPS.

The locations of several areas before and during their involution over the dorsal and ventral blastoporal lips are shown in Fig. T. At the blastula stage all of these areas are located on the surface. At the dorsal lip gastrula stage areas 1-5 have undergone involution over the *dorsal* blastoporal lip. At the yolk plug gastrula stage areas 1-8 have undergone involution over the *dorsal* blastoporal lip and areas 26 and 27 have undergone involution over the *ventral* blastoporal lip. The remaining areas will undergo involution during subsequent development. In a similar manner other cellular areas undergo involution over the lateral blastoporal lips.

Figure S shows the locations of several prospective areas some of which undergo involution during gastrulation. Some of the cells of the PROSPECTIVE ENDODERM and all of the cells of the PROSPECTIVE HEAD MESENCHYME and PROSPECTIVE NOTOCHORD involute over the *dorsal* blastoporal lip. Likewise, some of the cells of the PROSPECTIVE ENDODERM, some of the cells of the PROSPECTIVE LATERAL PLATE MESODERM, and all of the cells of the PROSPECTIVE SEGMENTAL PLATE MESODERM involute over the *lateral* lips of the blastopore. The remaining cells of the PROSPECTIVE LATERAL PLATE MESODERM as well as some of the cells of the PROSPECTIVE ENDODERM involute over the *ventral* blastoporal lip. Only a relatively small number of prospective endodermal cells undergo involution over the blastoporal lips. Most prospective endodermal cells remain relatively stationary during gastrulation stages forming the yolk plug and floor of the archenteron (Fig. T). As areas involute, the PROSPECTIVE NEURAL PLATE and PROSPECTIVE ECTODERM undergo spreading or EPIBOLY towards the blastopore to replace areas that have moved into the interior of the gastrula.

Obtain a slide containing sagittal sections of a dorsal lip gastrula and select a section closely resembling Plate 38, Fig. 2. Identify the BLASTOPORE, DORSAL BLASTOPORAL LIP, and BLASTOCOEL. The blastopore represents the future *caudal* end. The future *cranial* end lies directly opposite. The blastopore opens into a narrow cavity, the primitive gut or ARCHENTERON. The floor of the archenteron is formed by yolk-filled endodermal cells. The cranial end of the archenteron roof is formed from endoderm since the first cells to involute over the dorsal and lateral blastoporal lips and thus contribute to the wall of the archenteron are the cells of the PROSPECTIVE ENDODERM (Fig. S). The remainder of the archenteron roof at the gastrula stage is formed by mesoderm which involutes over the dorsal and lateral blastoporal lips following involution of prospective endoderm. The mesoderm of the archenteron roof will *later* be covered by endoderm which migrates upward over the inner surface of the mesoderm (see below). Thus the archenteron is ultimately lined by endoderm. Note that the ectoderm in the frog consists of two layers: an OUTER (SUPERFICIAL) ECTODERMAL LAYER and an INNER (DEEP) ECTODERMAL LAYER.

Obtain a slide containing sagittal sections of a yolk plug gastrula and select a section closely resembling Plate 38, Fig. 3. Identify the YOLK PLUG, a protruding mass of yolk-filled endodermal cells located between the very prominent DORSAL BLASTOPORAL LIP above and the less prominent VENTRAL BLASTOPORAL LIP below. Also identify the ARCHENTERON located above (dorsal to) the mass of poorly defined endodermal cells forming its floor. The yolk plug fills the entrance to the archenteron, the BLASTOPORE. Beneath the archenteron and separated from the latter by endodermal cells is the irregularly shaped BLASTOCOEL. This cavity is later squeezed out of existence. Note that the pigmented cortex now covers all cells except those of the yolk plug. Between the blastula stage and this advanced gastrula stage, the more rapidly dividing cells of the PROSPECTIVE ECTODERM and PROSPECTIVE NEURAL PLATE have undergone epiboly, growing down over the yolk-filled cells of the vegetal hemisphere. Meanwhile the large endodermal cells of the vegetal hemisphere have sunken inwards to some extent and become rearranged to form the floor and ventro-lateral walls of the archenteron. At the yolk plug gastrula stage most of the archenteron roof still consists of mesoderm.

E. NEURULATION

Aids: Plate 38, Figs. 4, 5; Plate 39, Figs. 1, 2; H, Fig. 56.

The ectodermal cells directly overlying the archenteron roof at the yolk plug gastrula stage constitute the NEURAL ECTODERM. The mesodermal cells of the archenteron roof *induce* the neural ectoderm to thicken as the NEURAL PLATE. With formation of the neural plate the gastrula is transformed into the NEURULA. The neural plate has the capacity to roll itself into a NEURAL TUBE, even when experimentally isolated from the neurula and grown in tissue culture.

Obtain a slide containing transverse sections of a neurula at the neural plate stage. Select a section closely resembling Plate 38, Figs. 4, 5 and identify the NEURAL PLATE. The roof of the ARCHENTERON at the neurula stage is formed by *endoderm*. Between the gastrula and neurula stage endoderm migrated upward over the inner surface of the mesoderm which previously formed the archenteron roof. Note that the dorsal mesoderm is now organized into a midline rod of cells, the NOTOCHORD, flanked by two bands of cells, the SEGMENTAL PLATE MESODERM. This latter mesoderm is the source of the SOMITES. The segmental plate mesoderm gradually merges laterally with the LATERAL PLATE MESODERM. The COELOM forms within this mesoderm.

Obtain a slide containing transverse sections of a neurula at the neural fold stage and select a section closely resembling Plate 39, Figs. 1, 2. Identify the paired NEURAL FOLDS, which have formed at the lateral margins of the neural plate, and the NEURAL GROOVE. The NOTOCHORD, SEGMENTAL PLATE MESODERM, and LATERAL PLATE MESODERM are more clearly defined at this stage, and the *endoderm* of the roof of the ARCHENTERON exists as a more distinct layer. During subsequent development the neural folds will fuse together in the dorsal midline to close the neural groove and thus establish the NEURAL TUBE.

F. 4-MM FROG EMBRYOS

1. Introduction. Aids: Fig. U; B, Fig. 234; M, Figs. 61, 62.

Fig. U shows the general shape of the body at this stage and the structures visible in a midsagittal section. Familiarize yourself with the spatial relationships of these structures before examining serial transverse sections.

2. Serial transverse sections. Aids: Plates 39-41; B, Figs. 235-241; M, Figs. 63-69.

Position your slide on the microscope stage so that when viewed through the microscope each section is oriented as in Plate 39, Fig. 3. Examine slides in antero-posterior sequence unless directed otherwise (see *Introduction*).

a. Ectodermal derivatives. The first few sections cut through the tip of the head. Identify the large PROSENCEPHALON (future TELENCEPHALON and DIENCEPHALON of the brain), which is enclosed by SKIN ECTODERM (Plate 39, Fig. 3). The skin ectoderm consists of two layers (not readily distinguishable from one another): the OUTER ECTODERMAL LAYER is heavily pigmented; the INNER ECTODERMAL LAYER contains far less pigment. Try to identify a region of thickened skin ectoderm on each side ventral to the prosencephalon. These thickenings are the NASAL (OLFACTORY) PLACODES, the rudiments of the NASAL CAVITIES. The nasal placodes are derived from only the *inner* ectodermal layer. The increased density of pigment granules at the periphery of these placodes is characteristic of invaginating ectodermal derivatives. The peripheral ends of invaginating cells become narrowed, thus concentrating their pigment content, whereas the inner ends of the cells enlarge. YOUNG NEURONS (formerly called NEUROBLASTS, a term no longer considered appropriate) *later* originate from these placodes and produce the AXONS of the OLFACTORY (I) CRANIAL NERVES, as well as the DENDRITES which function as OLFACTORY RECEPTORS (receptors for the sense of smell). Nasal placodes have not yet developed in some embryos. Also identify the PINEAL GLAND (EPIPHYSIS) which originated as a dorsal evagination from the prosen-

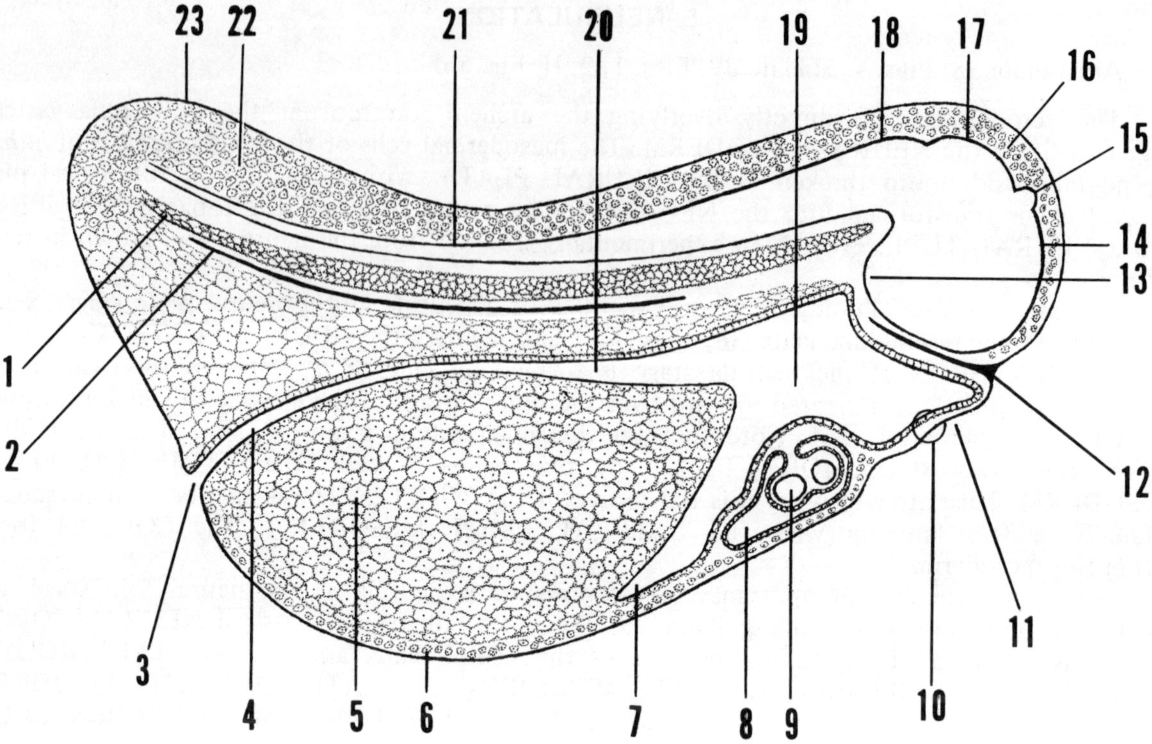

Fig. U. Schematic drawing of a midsagittal section of a 4-mm frog embryo.

1. Notochord
2. Subnotochordal rod
3. Anus
4. Hindgut
5. Yolk-filled endodermal cells
6. Lateral plate mesoderm
7. Liver rudiment
8. Pericardial cavity
9. Heart (ventricle)
10. Oral membrane
11. Stomodeum
12. Rudiment of the anterior pituitary gland
13. Infundibulum
14. Prosencephalon
15. Pineal gland
16. Head mesenchyme
17. Mesencephalon
18. Rhombencephalon
19. Pharynx
20. Midgut
21. Spinal cord
22. Neural crest cells
23. Dorsal fin

cephalon. Continue to trace sections posteriorly and identify the OPTIC CUPS (Plate 39, Fig. 4). Each optic cup is derived from a lateral evagination from the prosencephalon which secondarily invaginated at its blind end forming a double-layered optic cup. The thickened layer of the optic cup is the SENSORY RETINA, the thin layer is the PIGMENTED RETINA. The optic cups are connected to the prosencephalon by the OPTIC STALKS. Note that the *inner* ectodermal layer adjacent to the sensory retina is thickened as the LENS PLACODE (Plate 40, Fig. 1). The *outer* pigmented layer of the ectoderm will *later* be *induced* by the lens to form the transparent CORNEAL EPITHELIUM. At the level of the optic cups sections cut across the prosencephalon ventrally continuous with the MESENCEPHALON dorsally.

As sections are traced posteriorly the brain constricts into two separate parts (Plate 40, Fig. 2). The ventral part is the INFUNDIBULUM, source of the POSTERIOR PITUITARY GLAND (NEUROHYPOPHYSIS). The dorsal part is the RHOMBENCEPHALON (future METENCEPHALON and MYELENCEPHALON of the brain). At about this level identify a solid ectodermal rod just ventral to the infundibulum. This is the RUDIMENT OF THE ANTERIOR PITUITARY GLAND. Reverse direction and trace sections *anteriorly*. The rudiment of the anterior pituitary gland is continuous with an ectodermal invagination, the STOMODEUM, from which it originates as an outgrowth. (See Plate 40, Fig. 1; Plate 39, Fig. 4 for the location of the stomodeum. The rudiments of the anterior pituitary gland and the stomodeum are continuous at a section level between those illustrated by Plate 39, Figs. 4, 3.) Reverse direction and trace sections *posteriorly*. To either side of the stomodeum and extending through many sections identify prominent ectodermal thickenings (Plate 40, Fig. 1), the ADHESIVE GLANDS (VENTRAL SUCKERS).

Return to the level where the infundibulum first appears (Plate 40, Fig. 2) and trace sections posteriorly. As the optic cups fade from view, identify an accumulation of ectodermal cells (NEURAL CREST CELLS) just above each of them. These are the SEMILUNAR GANGLIA of the TRIGEMINAL (V) CRANIAL NERVES (Plate 40, Fig. 2). (*Later* cells derived from a thickening of the *inner* ectodermal layer on each side, an EPIBRANCHIAL PLACODE, will also contribute to these ganglia.) Continue to trace sections posteriorly and identify the paired AUDITORY VESICLES lying ventro-lateral to the rhombencephalon (Plate 40, Fig. 3). These vesicles originated from thickenings of the inner ectodermal layer which subsequently invaginated. The auditory vesicles later differentiate into the INNER EARS.

Continue to trace sections posteriorly and note that the neural tube gradually narrows as the SPINAL CORD (Plate 41, Fig. 1). Examine the spinal cord with high magnification. Note its characteristic thin ROOF and FLOOR PLATES and its thick lateral walls. At about this level identify pigmented or unpigmented cells lying dorsal to the spinal cord and beneath the skin ectoderm; these are NEURAL CREST CELLS (Plate 41, Fig. 1). As you approach the final sections in your set, identify the DORSAL FIN above the spinal cord (Plate 41, Figs. 2-5). The loose cells forming its core are NEURAL CREST CELLS, which *induce* formation of this structure. Note that the body narrows progressively at caudal levels and identify the VENTRAL FIN at the caudal end of the body (Plate 41, Fig. 5). A few sections *anterior* to the ventral fin identify the PROCTODEUM, a ventral invagination of skin ectoderm (Plate 41, Fig. 4).

b. Endodermal derivatives. Return to the level of the nasal placodes (Plate 39, Fig. 3) and trace sections posteriorly. Identify a small cavity lying beneath the prosencephalon (Plate 40, Fig. 1). This is the cranial end of the FOREGUT whose walls are formed from endoderm. The stomodeum has invaginated towards the foregut and the stomodeal ectoderm and the foregut endoderm are in contact as the ORAL MEMBRANE. This membrane *later* ruptures to form the MOUTH OPENING. The outer part of the mouth is therefore lined by stomodeal ectoderm, the inner part by foregut endoderm.

As sections are traced posteriorly the foregut expands as the PHARYNX (Plate 40, Fig. 3). (In a few embryos the lateral walls of the pharynx may contact the skin ectoderm which invaginates slightly to meet them. Such localized pharyngeal expansions are the PHARYNGEAL POUCHES.) Continue tracing sections posteriorly. The foregut narrows somewhat and then forms a prominent ventral evagination, the LIVER RUDIMENT (Plate 40, Fig. 4). The level of the foregut which is continuous ventrally with the liver rudiment is the DUODENUM. A few sections more posteriorly the liver rudiment separates from the duodenum and then fades out. The remaining portion of the gut is the MIDGUT (Plate 41, Figs. 1, 2). It contains a small cavity bounded dorsally by a thin layer of endodermal cells and ventrally by a large mass of yolk-filled endodermal cells. Continue to trace sections posteriorly following the midgut. It gradually moves ventrally and enlarges somewhat as the HINDGUT (Plate 41, Fig. 3). (In some embryos the endoderm of the hindgut fuses with the ectoderm of the proctodeum in more posterior sections to form the CLOACAL MEMBRANE. The cloacal membrane ultimately ruptures to form the ANUS.)

c. Mesodermal derivatives. Return to the level where the infundibulum first appears (Plate 40, Fig. 2) and trace sections posteriorly. Identify a group of mesodermal cells, the NOTOCHORD, lying beneath the rhombencephalon (Plate 40, Fig. 3). The vacuolated condition of the notochord is responsible for its rigidity. Therefore this structure can serve as a longitudinal supporting rod in young embryos. Amphibian embryos will not elongate properly in the absence of the notochord so this structure does more than merely support the developing embryo. Quickly trace sections posteriorly, noting the changes that the notochord undergoes. It is smaller at caudal levels than at more cranial levels and vacuolization is progressively reduced towards the caudal end (cf. Plate 41, Figs. 1-5). The caudal end of the notochord is less differentiated at this stage because the notochord differentiates in cranio-caudal sequence. In posterior sections identify the SOMITES (Plate 41, Figs. 2, 3), paired blocks of mesoderm lying ventro-lateral to the spinal cord. Also identify a small cluster of mesodermal cells lying beneath the notochord in caudal regions, the SUBNOTOCHORDAL ROD (Plate 41, Figs. 2, 4). Its developmental significance is unknown.

Return to the level at which the liver rudiment is continuous with the duodenum (Plate 40, Fig. 4) and trace sections posteriorly. The mesoderm ventro-lateral to the somites is usually in the form of more or less distinct epithelial vesicles (Plate 41, Fig. 1). The round vesicle on each side farthest from the spinal cord is the PRONEPHRIC DUCT. Just above the latter, and often continuous with it, are the PRONEPHRIC TUBULES. The pronephric duct and tubules on each side constitute the PRONEPHRIC KIDNEY, which is functional in amphibian larvae. The pronephric kidney when well formed causes the body to bulge laterally as the PRONEPHRIC RIDGE.

Ventral to each pronephric kidney between the skin ectoderm and the endoderm lies the LATERAL PLATE MESODERM (often difficult to identify as a distinct area). This mesoderm is in the process of splitting into an outer layer of SOMATIC MESODERM adjacent to the skin ectoderm, and an inner layer of SPLANCHNIC MESODERM adjacent to the endoderm. If somatic and splanchnic mesoderm have formed, identify a space (or a series of small spaces) between them, the COELOM.

Return to the level at which the foregut first appears (Plate 40, Fig. 1) and trace sections posteriorly. Identify the BULBUS CORDIS region of the developing heart lying beneath the pharynx (Plate 40, Fig. 3). In more posterior sections the heart enlarges as the VENTRICLE (Plate 40, Fig. 4). Note that the heart consists of an inner layer, the ENDOCARDIUM, surrounded by an outer, thicker layer, the EPI-MYOCARDIUM. Both of these layers are derived from *splanchnic* mesoderm. Identify a thin layer of cells enclosing the heart, the PARIETAL PERICARDIUM, derived from *somatic* mesoderm. The parietal pericardium is usually separated from the skin ectoderm by a shrinkage space. The space between the heart and the parietal pericardium is the PERICARDIAL CAVITY. The heart is suspended within the pericardial cavity by a dorsal bridge of *splanchnic* mesoderm, the DORSAL MESOCARDIUM.

The major blood vessels are in early stages of formation at this time and they are thus difficult to identify with certainty in most embryos. In some embryos two major blood vessels can be identified. The FIRST AORTIC ARCHES primarily lie ventro-lateral to the pharynx (Plate 40, Fig. 3). (They will soon establish connections with the bulbus cordis via a pair of blood vessels which will lie ventral to the pharynx, the VENTRAL AORTAE. These latter vessels can sometimes be identified at this stage.) The dorsal end of each first aortic arch is continuous with a blood vessel lying dorso-lateral to the pharynx, the DORSAL AORTA (Plate 40, Fig. 3). The dorsal aortae extend caudad and fade out at about the level of the midgut (Plate 41, Fig. 1).

3. Summary of germ layer derivatives present at the 4-mm stage.

Ectoderm		
adhesive glands	nasal placodes	rhombencephalon
auditory vesicles	neural crest cells	rudiment of the anterior pituitary gland
corneal epithelium	optic cups	semilunar ganglia
dorsal fin	optic stalks	stomodeum
infundibulum	pineal gland	spinal cord
lens placode	proctodeum	ventral fin
mesencephalon	prosencephalon	

Mesoderm	*Endoderm*	*Ectoderm and Endoderm*
bulbus cordis	duodenum	cloacal membrane
dorsal aortae	foregut	oral membrane
dorsal mesocardium	hindgut	
first aortic arches	liver rudiment	
notochord	midgut	
parietal pericardium	pharynx	
pronephric kidneys		
somites		
subnotochordal rod		
ventricle		

PLATE 1

33-HOUR CHICK EMBRYO WHOLE MOUNTS

(Figures 1-4 dorsal views; Figure 5 ventral view)

1. Area vasculosa
2. Blood islands
3. Somites
4. Spinal cord
5. Segmental plates
6. Neural folds
7. Region of the primitive streak
8. Infundibulum
9. Proamnion
10. Mesencephalon
11. Caudal extent of the subcephalic pocket
12. Metencephalon
13. Neuromeres of myelencephalon
14. 5th intersomitic furrow
15. Notochord beneath broadened neural groove
16. Optic vesicle(s)
17. Notochord beneath mesencephalon
18. First somite
19. Area pellucida
20. Continuity between a neural fold (upper arrow) and primitive ridge (lower arrow)
21. Cranial neuropore
22. Skin ectoderm
23. Head mesenchyme
24. Foregut
25. Ventricle
26. Sino-atrial region
27. Cranial intestinal portal
28. Bulbus cordis
29. Continuity between sino-atrial region and vitelline vein

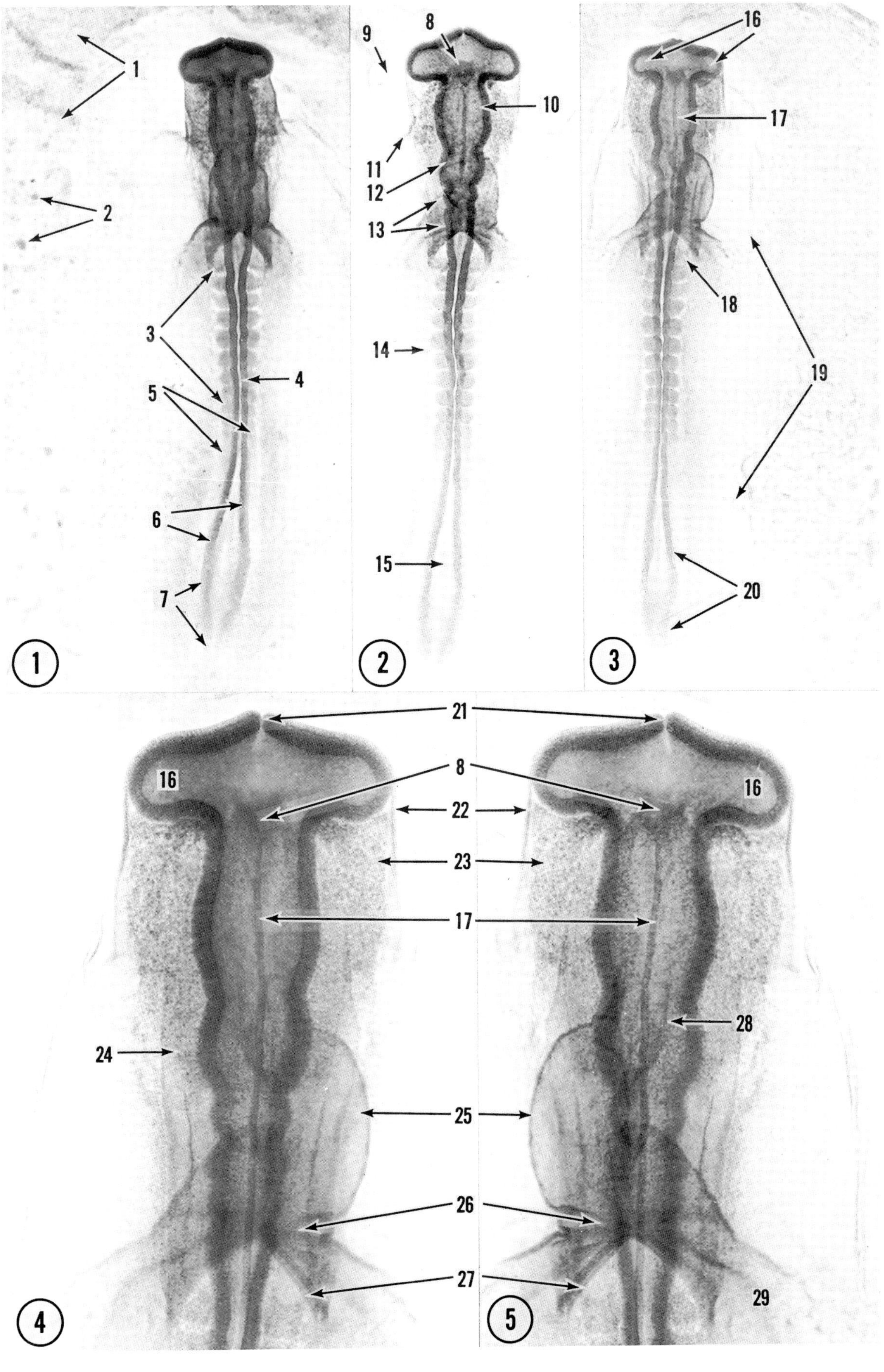
1
2
3
4
5
6
7
8
9
10
11
12
13
14
15
16
17
18
19
20
21
22
23
24
25
26
27
28
29
1
2
3
4
5

PLATE 2

33-HOUR CHICK EMBRYO SERIAL TRANSVERSE SECTIONS

1. Diencephalon
2. Future lens ectoderm
3. Optic vesicle
4. Infundibulum
5. Foregut
6. Subcephalic pocket
7. Proamnion
8. Head mesenchyme
9. Dorsal aorta
10. First aortic arch
11. Ventral aorta
12. Oral membrane
13. Notochord
14. Mesencephalon
15. Coelom
16. Splanchnic mesoderm
17. Endoderm
18. Bulbus cordis
19. Somatic mesoderm
20. Skin ectoderm
21. Somatopleure
22. Metencephalon

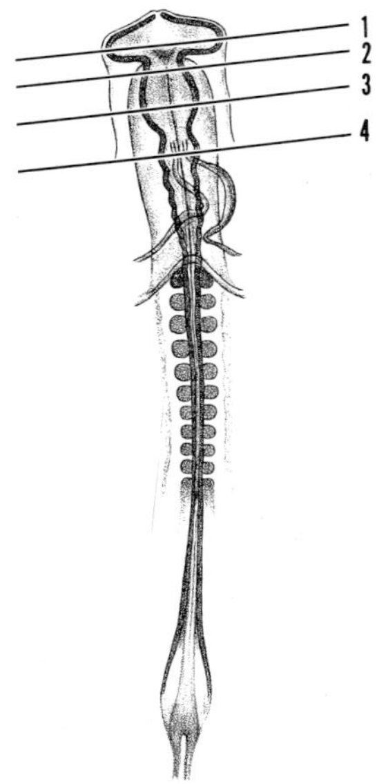

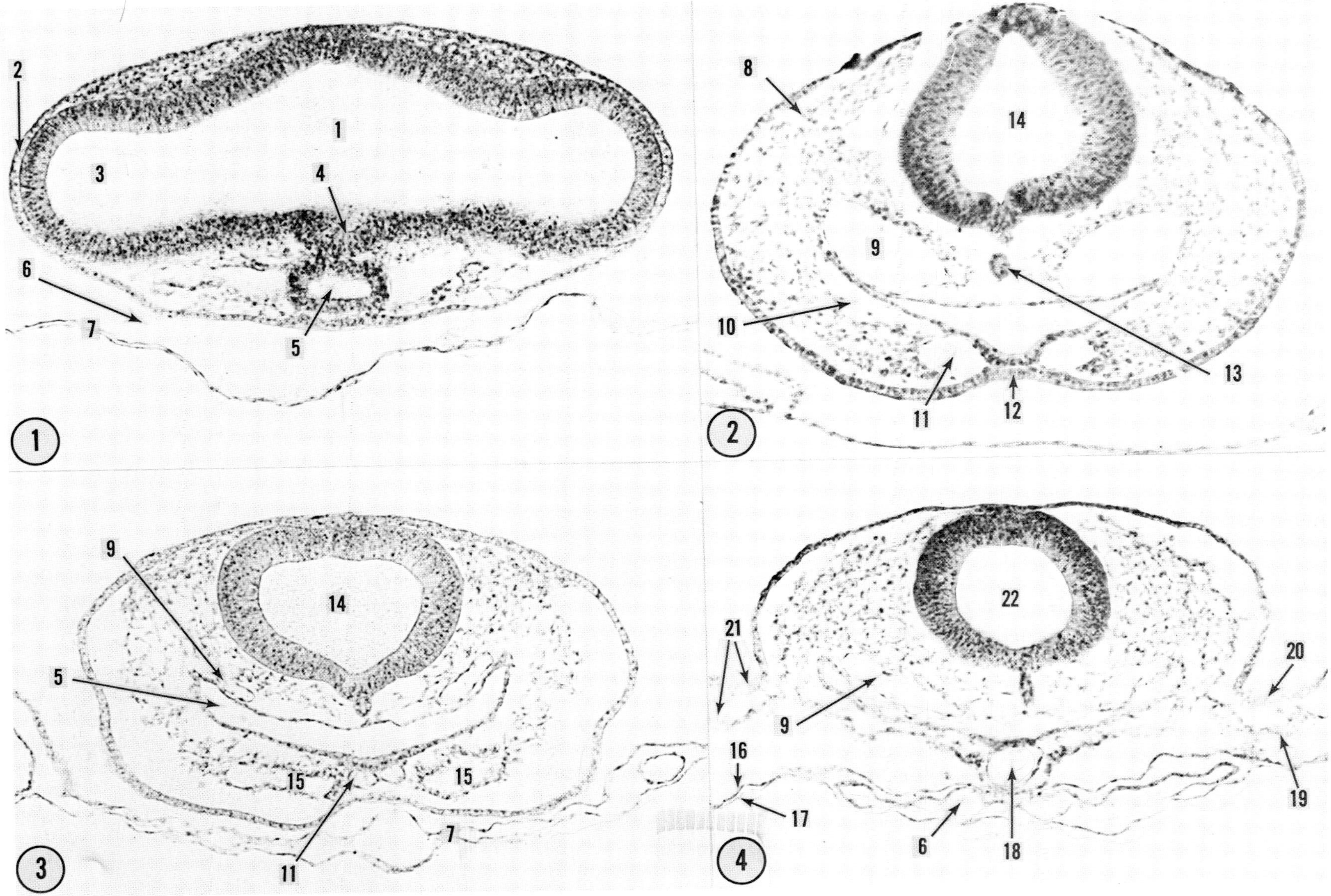

1
2
3
4
5
6
7
8
9
10
11
12
13
14
15
16
17
18
19
20
21
22

PLATE 3

33-HOUR CHICK EMBRYO SERIAL TRANSVERSE SECTIONS AND SAGITTAL SECTION

(Figures 1-5 serial transverse sections; Figure 6 midsagittal section)

1. Skin ectoderm
2. Foregut
3. Ventricle
4. Pericardial cavity
5. Epi-myocardium
6. Endocardium
7. Thyroid rudiment
8. Dorsal aorta
9. Myelencephalon
10. Auditory placode
11. Junction between somatic and splanchnic mesoderm
12. Sino-atrial region
13. Roof plate
14. Spinal cord
15. Floor plate
16. Pronephric cord
17. Splanchnopleure
18. Notochord
19. Endoderm
20. Splanchnic mesoderm
21. Somite
22. Neural groove
23. Neural folds
24. Segmental plate
25. Primitive groove
26. Primitive ridges
27. Primitive streak
28. Infundibulum
29. Mesencephalon
30. Metencephalon
31. Cranial intestinal portal
32. Bulbus cordis
33. Proamnion
34. Oral membrane
35. Subcephalic pocket
36. Prosencephalon

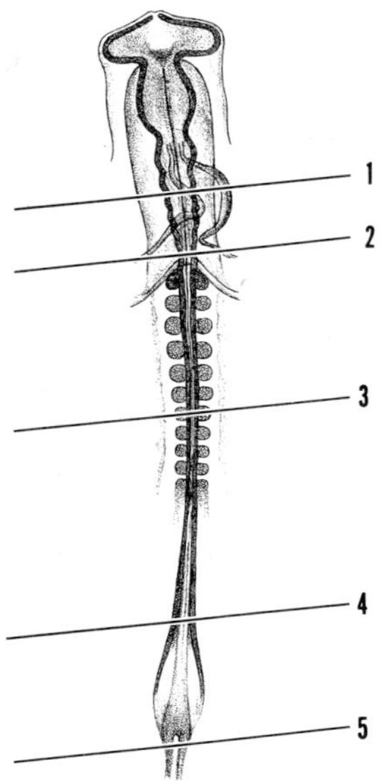

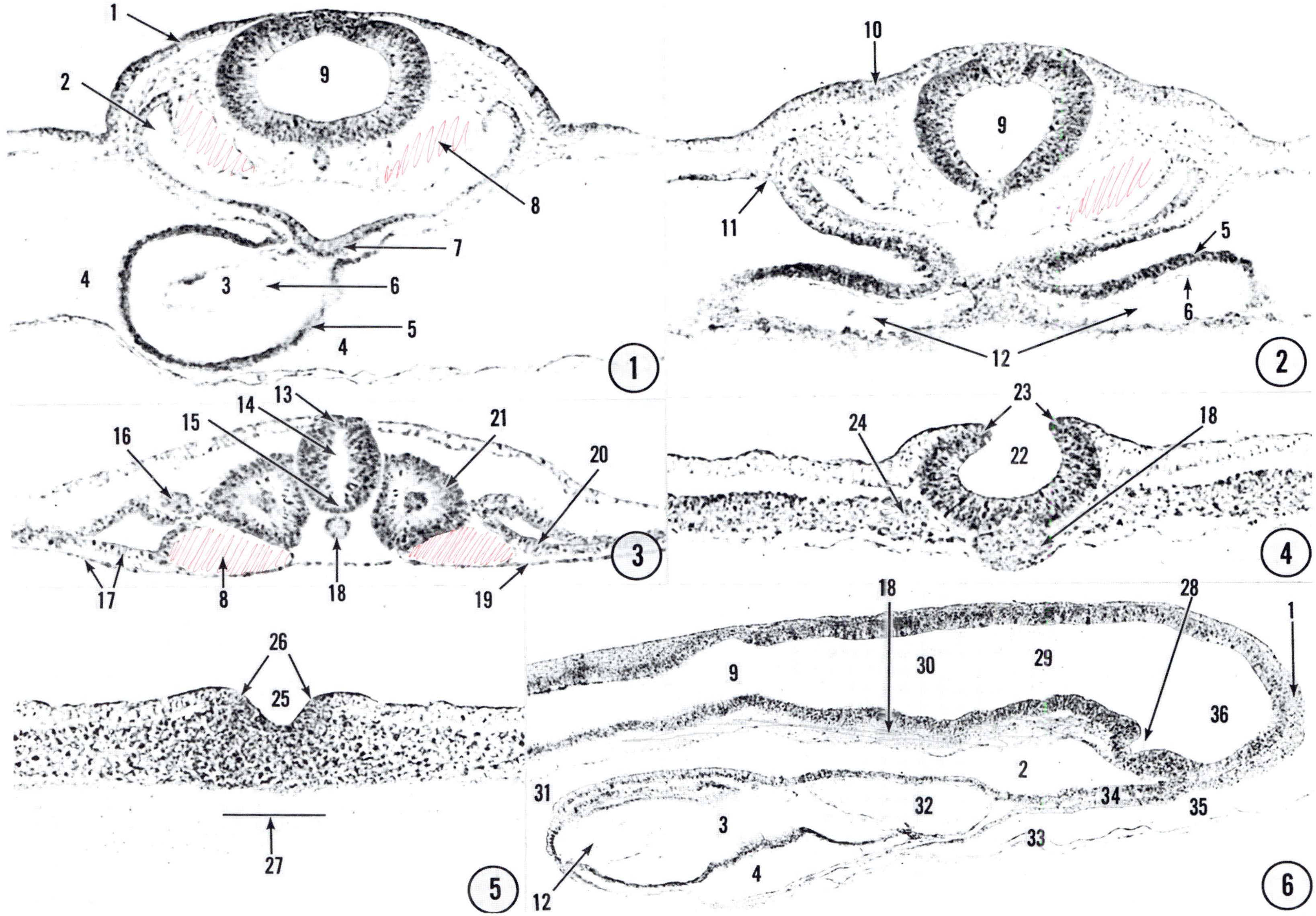
1
2
3
4
5
6
7
8
9
10
11
12
13
14
15
16
17
18
19
20
21
22
23
24
25
26
27
28
29
30
31
32
33
34
35
36

PLATE 4

18-HOUR CHICK EMBRYO WHOLE MOUNTS

FIGURE 1 Definitive primitive streak stage

FIGURES 2, 3 Head process stage

FIGURE 4 Early head fold stage

1. Area pellucida
2. Primitive streak
3. Head process
4. Primitive knot
5. Primitive groove
6. Primitive ridge
7. Primitive pit
8. Proamnion
9. Head fold of the body

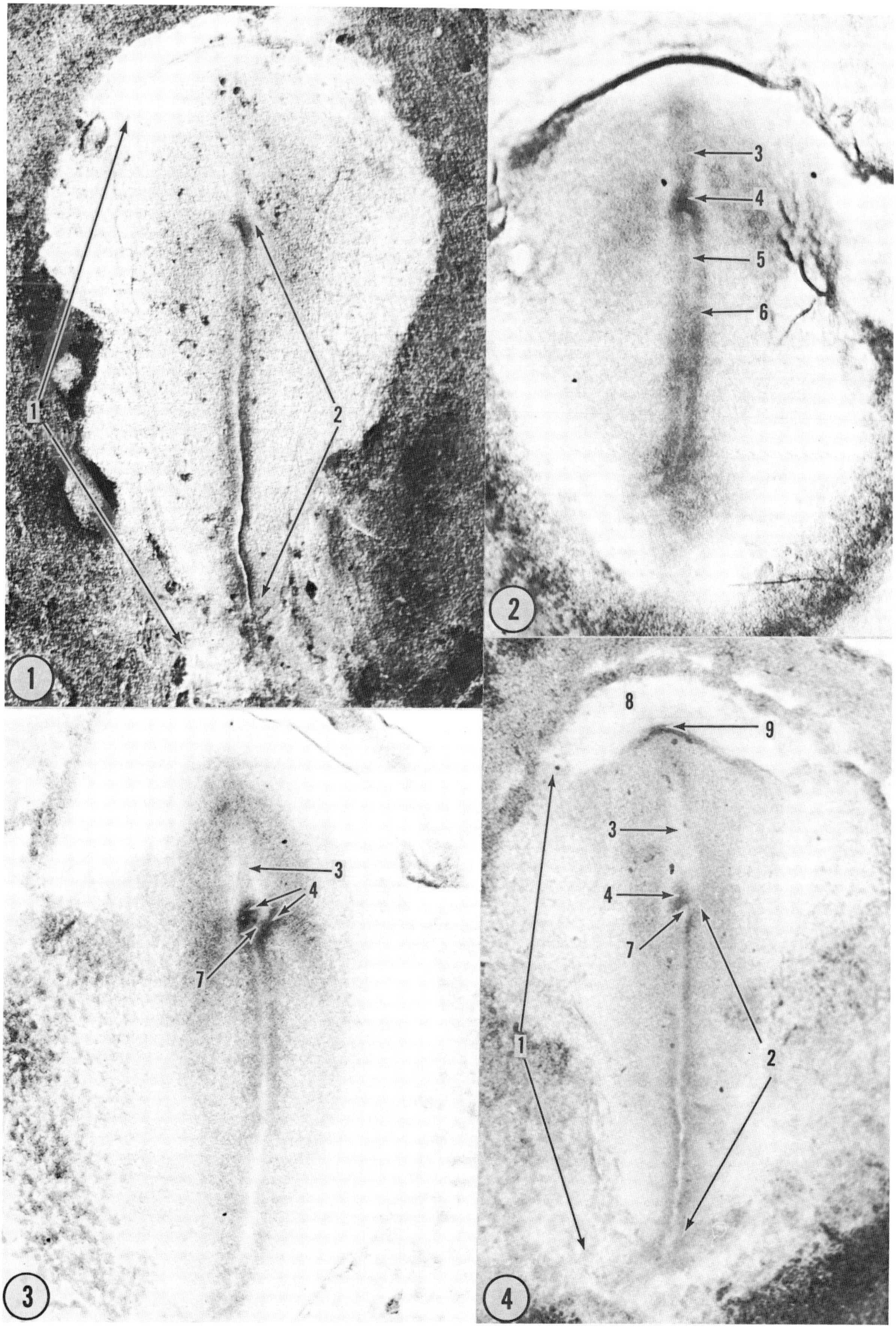
1
2
1
2
3
4
5
6
3
4
7
3
8
9
3
4
7
1
2
4

PLATE 5

24-HOUR CHICK EMBRYO WHOLE MOUNTS

(Figures 3, 4 enlargements of embryo illustrated in Figure 2)

1. Proamnion
2. Caudal extent of the subcephalic pocket
3. Neural folds
4. Somites
5. Cranial neuropore
6. Primitive streak
7. Neural tube
8. Cranial intestinal portal
9. Area opaca
10. Intersomitic furrows
11. Notochord
12. Area pellucida
13. Skin ectoderm
14. Head mesenchyme
15. Lateral margin of foregut
16. Neural groove
17. Second somite
18. Third intersomitic furrow
19. Segmental plate

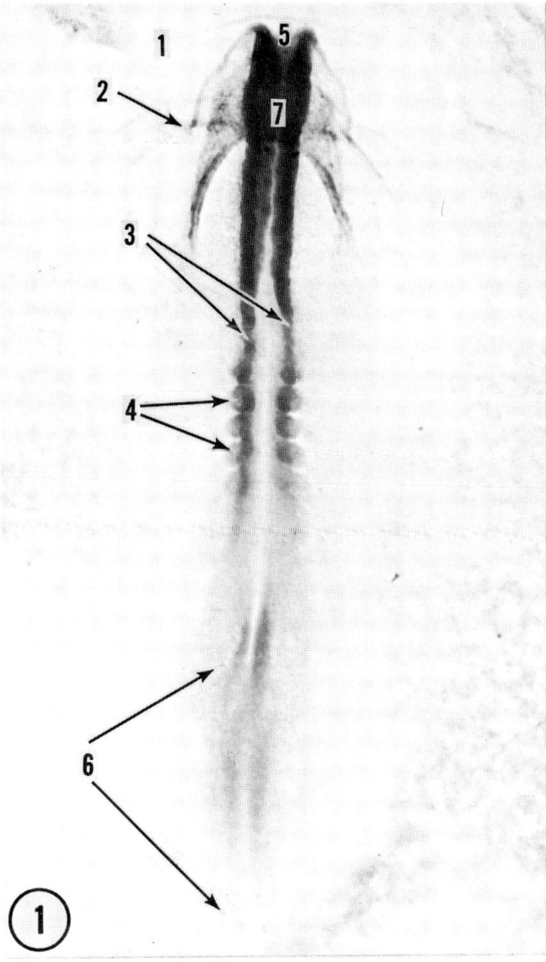
1
2
5
7
3
4
6
1

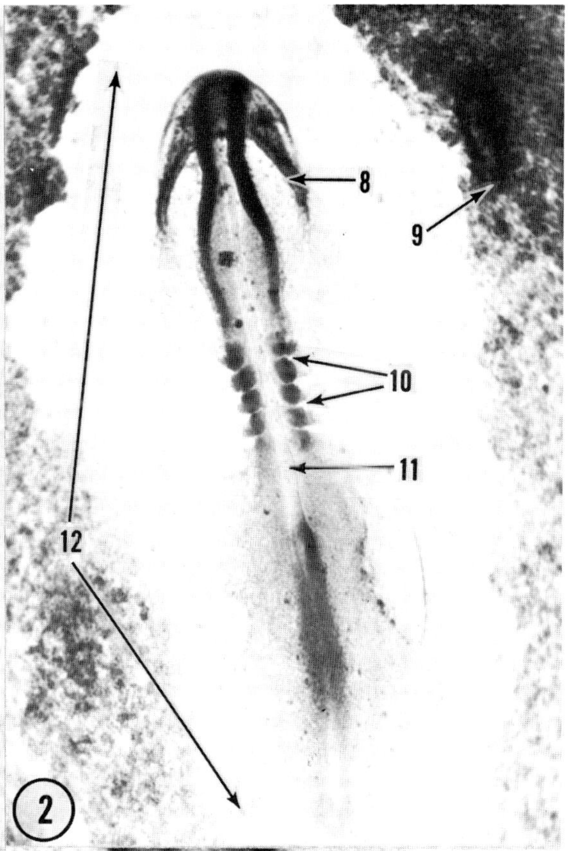
8
9
10
11
12
2

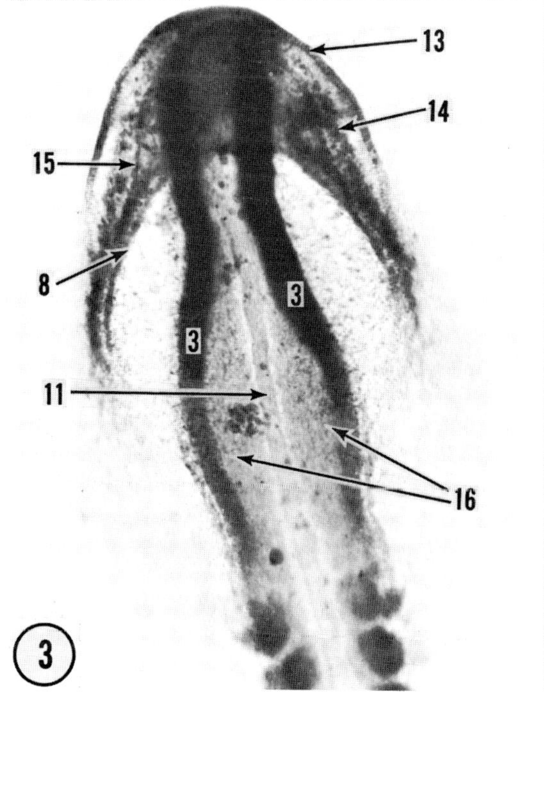
13
14
15
8
3
3
11
16
3

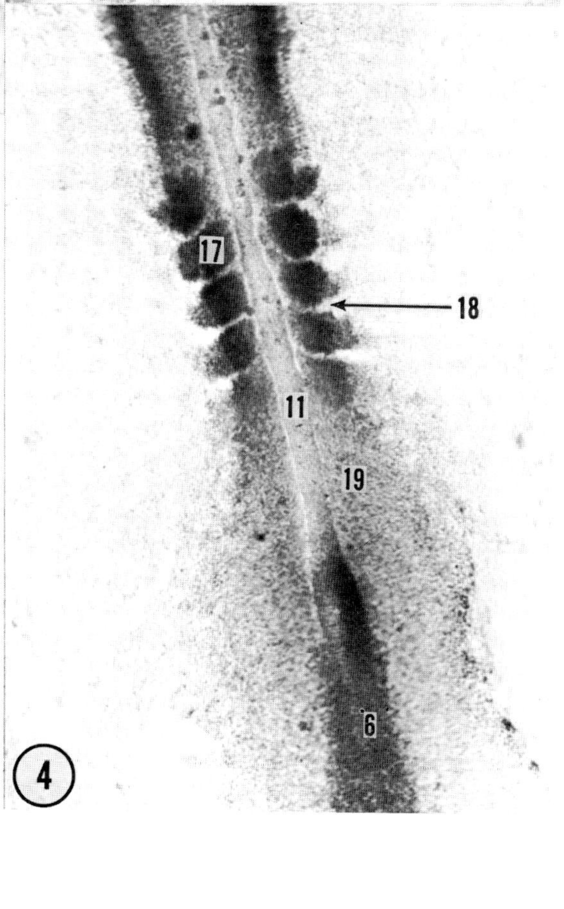
17
18
11
19
6
4

PLATE 6

24-HOUR CHICK EMBRYO SERIAL TRANSVERSE SECTIONS AND SAGITTAL SECTIONS

FIGURES 1-4 Serial transverse sections

FIGURE 5 Midsagittal section

FIGURE 6 Sagittal section just lateral to the midline

(Figure 2 unlabeled arrows indicate lateral body fold)

1. Neural tube
2. Skin ectoderm
3. Head mesenchyme
4. Foregut
5. Coelom
6. Proamnion
7. Subcephalic pocket
8. Notochord
9. Somatopleure
10. Splanchnopleure
11. Rudiment of pericardial cavity
12. Cranial intestinal portal
13. Cardiac primordia
14. Endoderm
15. Splanchnic mesoderm
16. Somatic mesoderm
17. Neural groove
18. Neural ectoderm
19. Somite
20. Neural fold
21. Beginning coelom formation in lateral plate
22. Nephrotome
23. Neural plate
24. Segmental plate
25. Oral membrane
26. Cranial neuropore

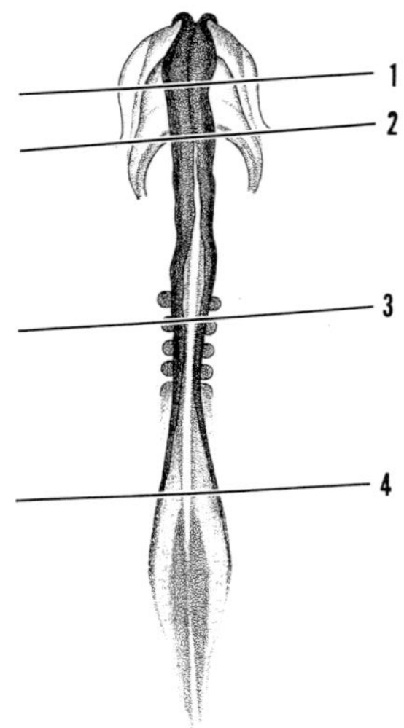

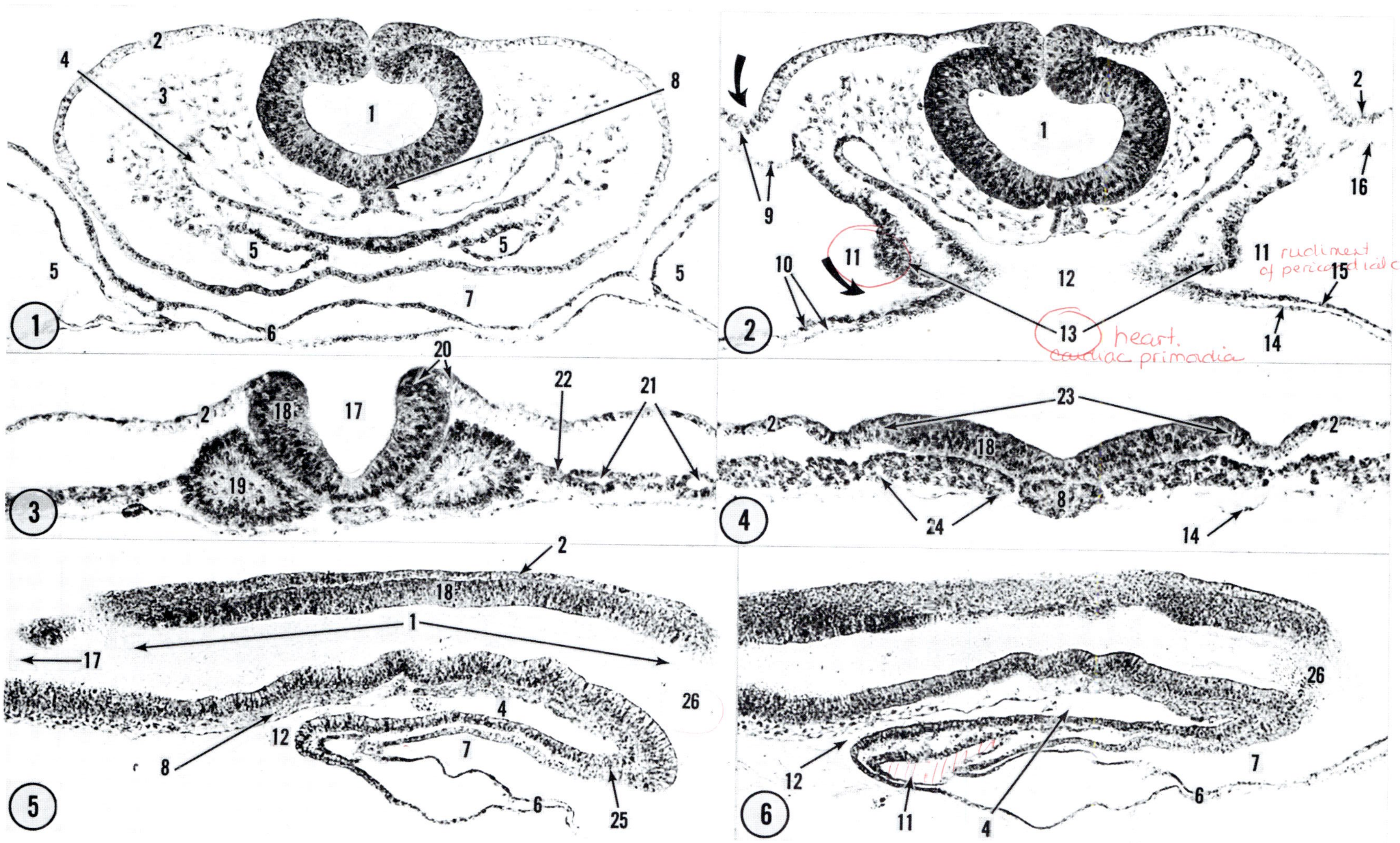
1
2
3
4
5
6
7
8
9
10
11
12
13
14
15
16
17
18
19
20
21
22
23
24
25
26

PLATE 7

48-HOUR CHICK EMBRYO WHOLE MOUNTS

(Figure 2 * indicates the cranial flexure)

1. Stomodeum
2. Optic cup
3. Maxillary process
4. Notochord
5. Isthmus
6. Mandibular process
7. Auditory vesicle
8. Second branchial groove
9. Third branchial groove
10. Caudal intestinal portal
11. Caudal boundary of body
12. Cranial intestinal portal
13. Cranial liver rudiment
14. Ventricle
15. Bulbus cordis
16. Optic fissure
17. Metencephalon
18. Roof plate of myelencephalon
19. First branchial groove
20. Second branchial arch
21. Third branchial arch
22. Boundary of amniotic folds
23. Spinal cord
24. Telencephalon
25. Foregut
26. Diencephalon
27. Mesencephalon
28. Myelencephalon
29. Somites
30. Vitelline artery
31. Tail bud

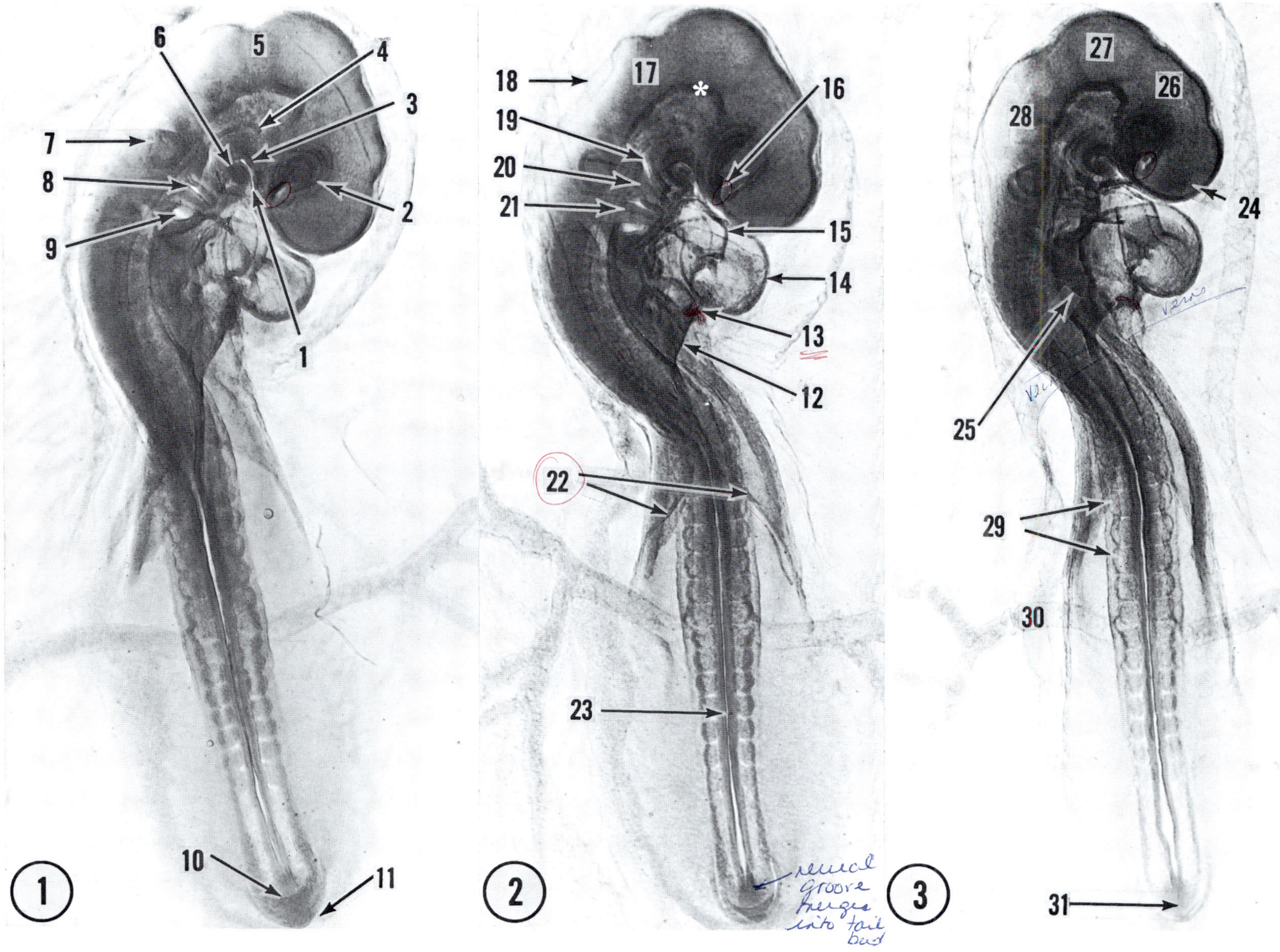
1
2
3
4
5
6
7
8
9
10
11
12
13
14
15
16
17
18
19
20
21
22
23
24
25
26
27
28
29
30
31
*

①

②

③

PLATE 8

48-HOUR CHICK EMBRYO WHOLE MOUNTS – INJECTED

(Figure 2 cranial half of embryo illustrated in Figure 1)

1. Right vitelline vein
2. Plexus of head blood vessels
3. Left vitelline vein
4. Plexus of vitelline blood vessels
5. Descending aorta
6. Dorsal aorta
7. Vitelline artery
8. Right vitelline vein entering sinus venosus
9. Sinus venosus
10. Common cardinal and left vitelline veins entering sinus venosus
11. Ventricle
12. Bulbus cordis
13. Atrium
14. Aortic sac
15. First aortic arch
16. Continuity between dorsal aorta and first aortic arch (dashed line)
17. Precardinal vein
18. Second aortic arch
19. Third aortic arch
20. Intersegmental arteries

1
2
3
4
5
6
7
8
9
10
11
12
13
14
15
16
17
18
19
20

PLATE 9

48-HOUR CHICK EMBRYO SERIAL TRANSVERSE SECTIONS

1. Myelencephalon
2. Metencephalon
3. Isthmus
4. Head mesenchyme
5. Mesencephalon
6. Diencephalon
7. Roof plate of myelencephalon
8. Semilunar ganglion
9. Notochord
10. Amnion
11. Yolk sac
12. Precardinal vein
13. Acoustico-facialis ganglion
14. Dorsal aorta
15. Foregut endoderm
16. Auditory vesicle
17. First pharyngeal pouch
18. First branchial groove
19. Pharynx
20. First aortic arch
21. Chorion
22. Rathke's pouch
23. Infundibulum
24. First closing plate

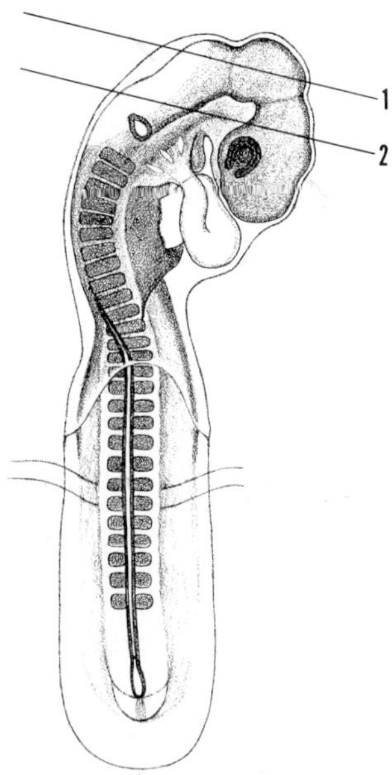

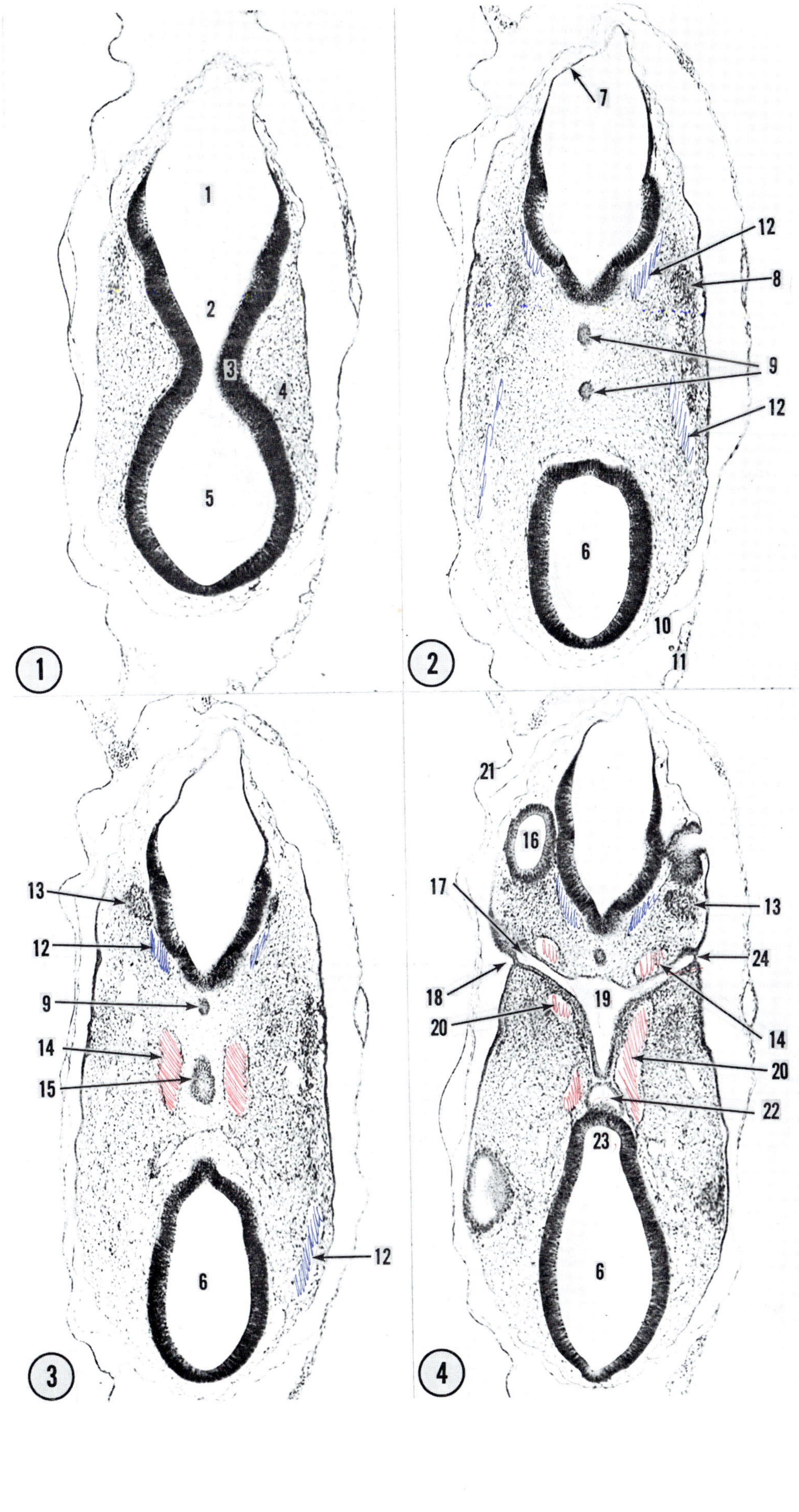

1
2
3
4
5
1
7
12
8
9
12
6
10
11
2
13
12
9
14
15
6
12
3
21
16
17
13
24
18
19
20
14
20
22
23
6
4

PLATE 10

48-HOUR CHICK EMBRYO SERIAL TRANSVERSE SECTIONS

1. Myelencephalon
2. Superior ganglion
3. Auditory vesicle
4. Acoustico-facialis ganglion
5. First pharyngeal pouch
6. Pharynx
7. First branchial groove
8. First aortic arch
9. Mandibular process
10. Maxillary process
11. Oral membrane
12. Stomodeum
13. Rathke's pouch (continuous with stomodeum)
14. Infundibulum
15. Optic cup
16. Epibranchial placode (contributes cells to facialis part of acoustico-facialis ganglion)
17. Opticoel
18. Precardinal vein
19. Notochord
20. Dorsal aorta
21. Second aortic arch
22. Pigmented retina
23. Continuity between skin ectoderm and lens vesicle
24. Sensory retina
25. Skin ectoderm
26. Second pharyngeal pouch
27. Aortic sac
28. Vitelline blood vessel
29. Second closing plate
30. Chorion
31. Amnion
32. Yolk sac
33. Optic stalk
34. Thyroid rudiment
35. Diencephalon

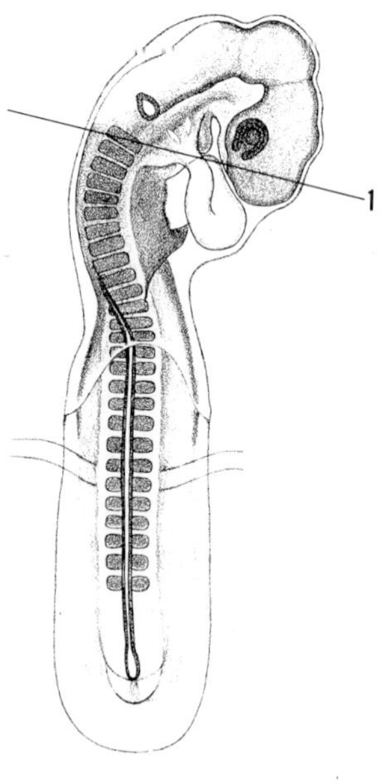

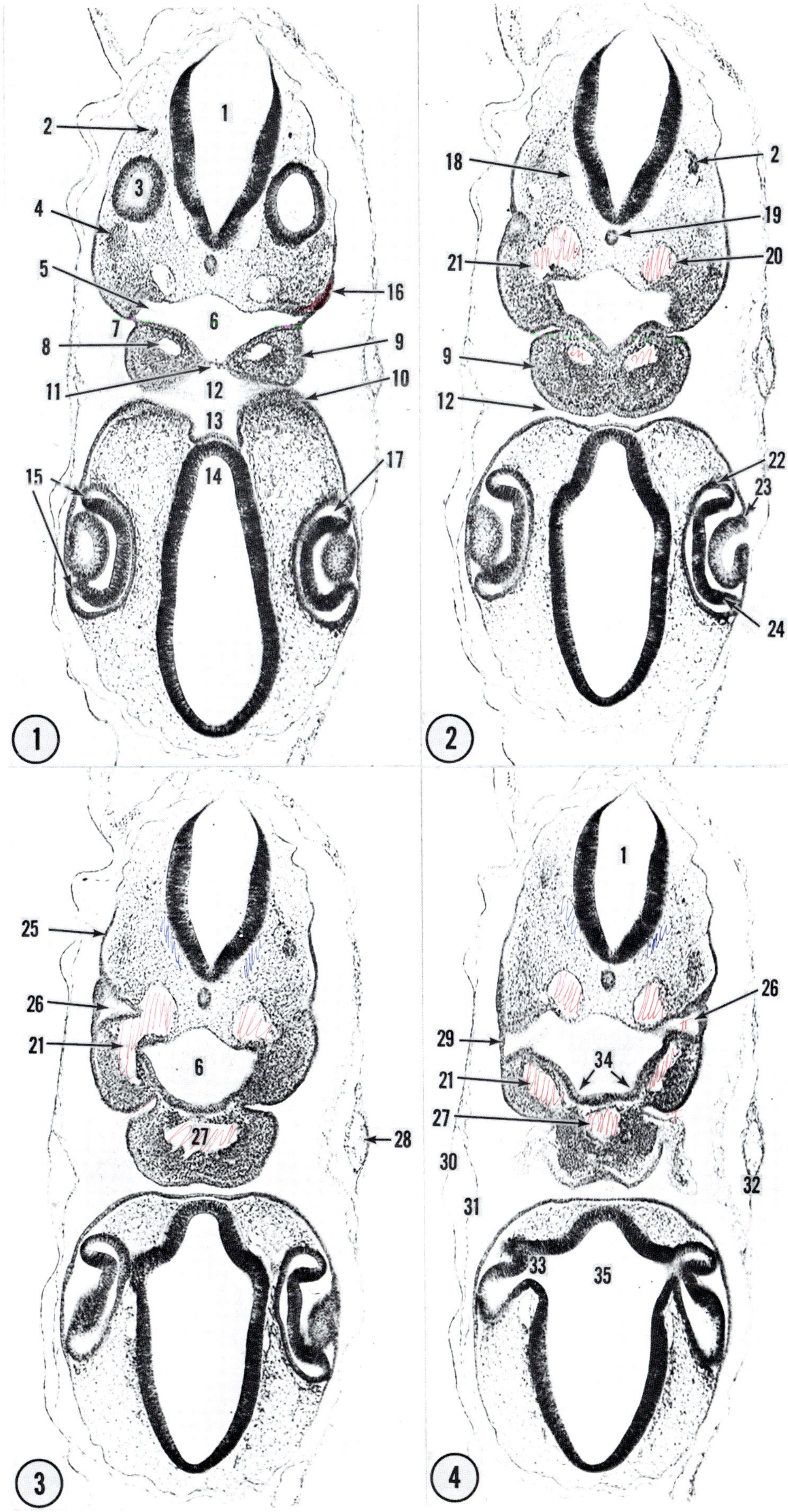
1
2
3
4
5
6
7
8
9
10
11
12
13
14
15
16
17
①
2
18
19
20
21
9
12
22
23
24
②
25
26
21
6
27
28
③
1
26
29
34
21
27
30
31
32
33
35
④

PLATE 11

48-HOUR CHICK EMBRYO SERIAL TRANSVERSE SECTIONS

1. Somite #1
2. Chorion
3. Precardinal vein
4. Dorsal aortae
5. Amniotic cavity
6. Third pharyngeal pouch
7. Third closing plate
8. Pharynx
9. Third aortic arch
10. Endocardium of bulbus cordis
11. Epi-myocardium of bulbus cordis
12. Aortic sac
13. Nasal placode
14. Telencephalon
15. Dermatome
16. Myotome
17. Sclerotome
18. Atrium
19. Descending aorta
20. Foregut
21. Amnion
22. Splanchnic mesoderm
23. Pericardial cavity
24. Bulbus cordis
25. Ventricle
26. Spinal cord
27. Pleural cavity
28. Postcardinal vein
29. Common cardinal vein
30. Beginnings of lung bud
31. Dorsal mesocardium
32. Sinus venosus
33. Vitelline blood vessel
34. Cranial liver rudiment

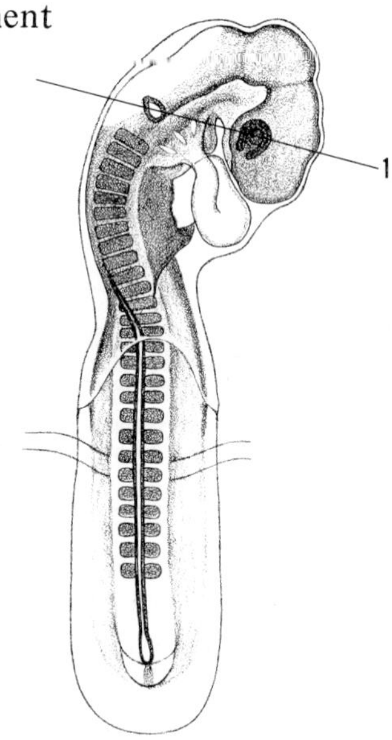

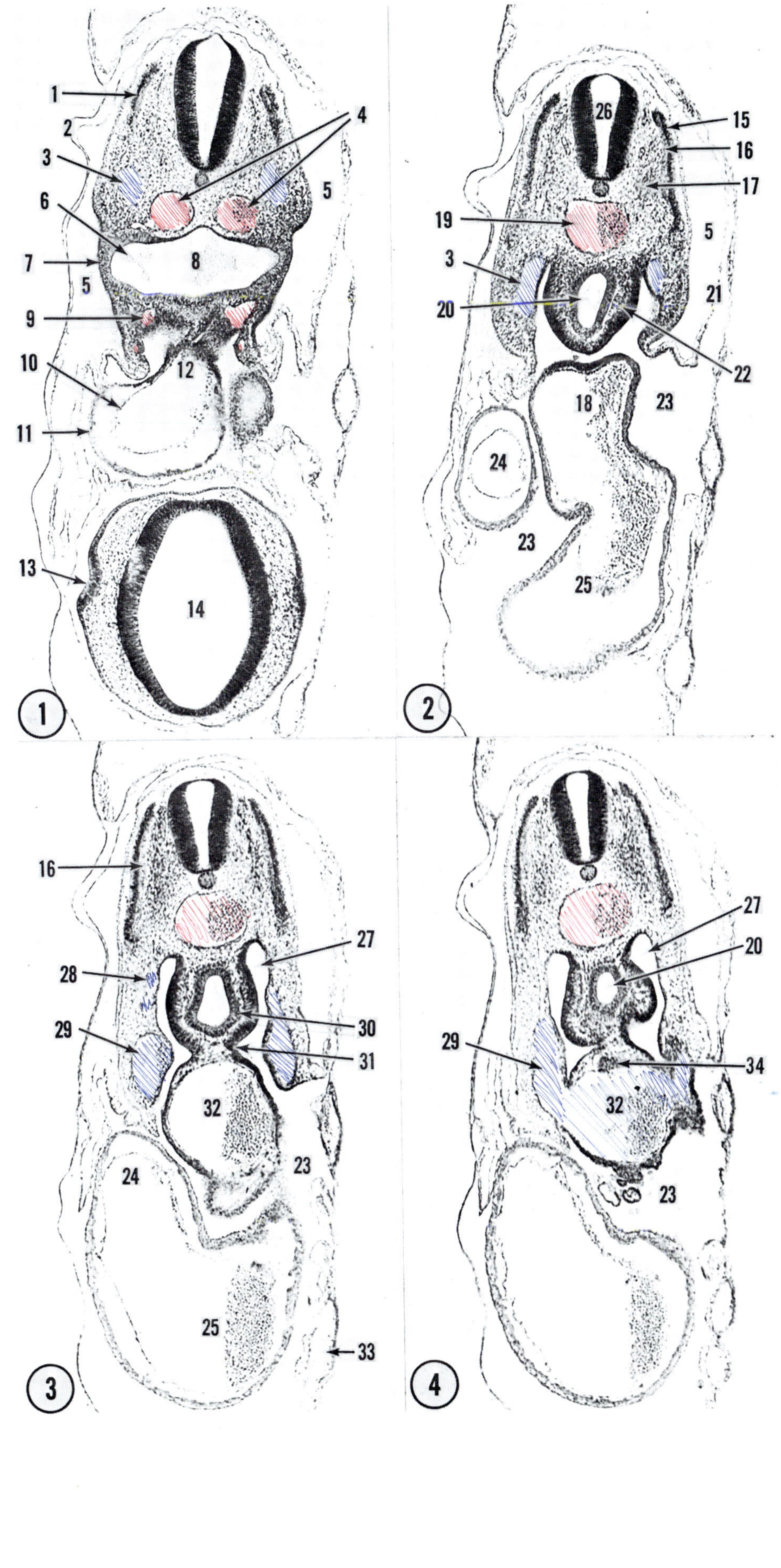
1
2
3
4
5
6
7
8
9
10
11
12
13
14
15
16
17
18
19
20
21
22
23
24
25
26
27
28
29
30
31
32
33
34

PLATE 12

48-HOUR CHICK EMBRYO SERIAL TRANSVERSE SECTIONS

1. Spinal cord
2. Notochord
3. Intersegmental artery
4. Descending aorta
5. Postcardinal vein
6. Duodenum
7. Cranial liver rudiment
8. Vitelline vein
9. Caudal liver rudiment
10. Rudiment of the peritoneal cavity
11. Mesonephric duct rudiment (cavitated)
12. Cranial intestinal portal
13. Yolk sac
14. Mesonephric tubule rudiment
15. Ectoderm of lateral amniotic fold
16. Somatic mesoderm of lateral amniotic fold
17. Chorion
18. Amnion
19. Amniotic cavity
20. Dorsal aortae
21. Extra-embryonic coelom
22. Mesonephric duct rudiment (noncavitated)
23. Sclerotome
24. Myotome
25. Dermatome
26. Vitelline blood vessel
27. Floor plate
28. Roof plate
29. Dorsal aorta
30. Vitelline artery
31. Ectoderm of lateral body fold
32. Somatic mesoderm of lateral body fold
33. Endoderm

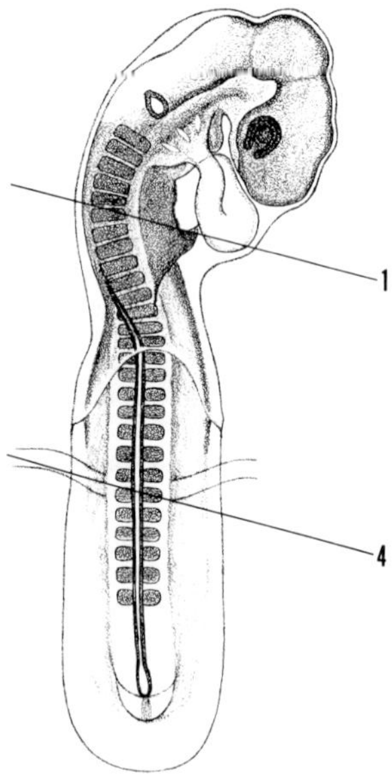

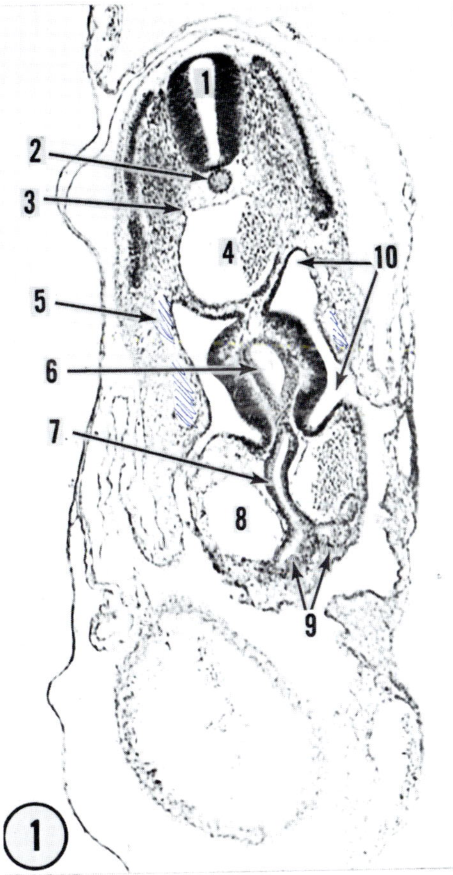
1
2
3
4
5
6
7
8
9
10
1

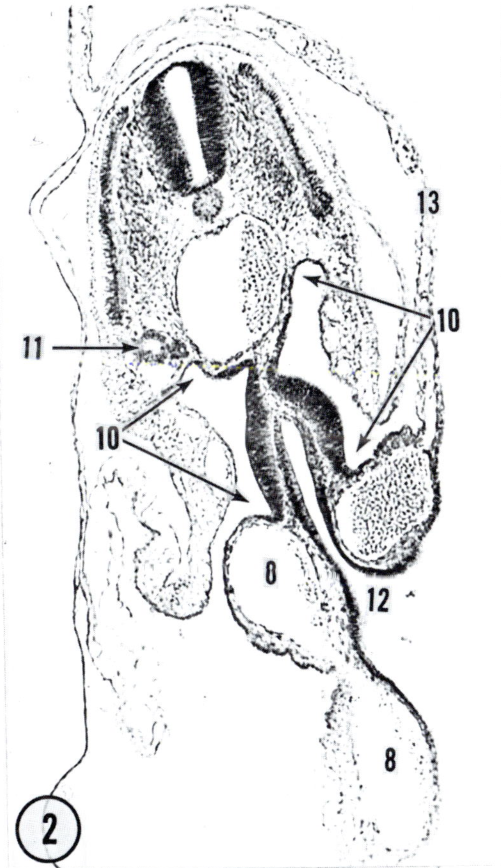
13
10
11
10
8
12
8
2

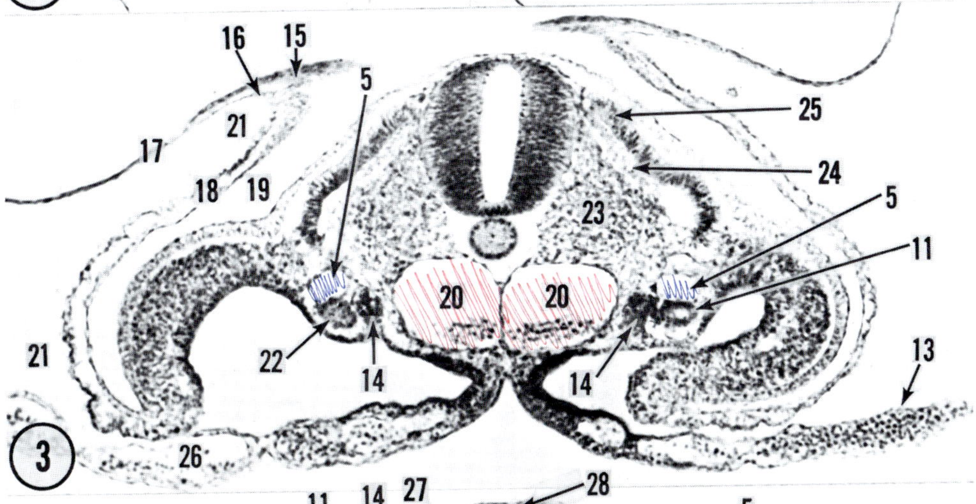
16
15
5
25
21
17
24
18
19
23
5
11
20
20
21
22
14
14
13
26
3

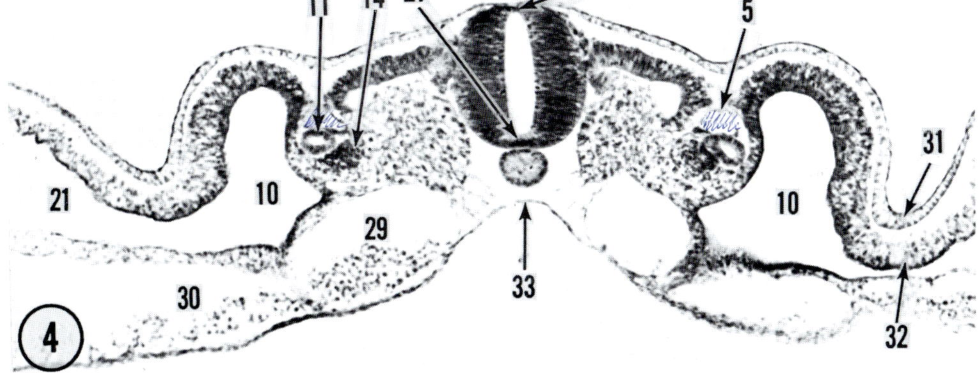
11
14
27
28
5
31
21
10
29
10
33
30
32
4

PLATE 13

48-HOUR CHICK EMBRYO SERIAL TRANSVERSE SECTIONS AND SAGITTAL SECTIONS

FIGURES 1, 2 Serial transverse sections

FIGURE 3 Midsagittal section

FIGURE 4 Parasagittal section at level of the optic cup

1. Tail bud
2. Somatopleure
3. Vitelline blood vessel
4. Caudal intestinal portal
5. Skin ectoderm
6. Extra-embryonic coelom
7. Splanchnopleure (of yolk sac)
8. Allantois rudiment
9. Roof plate of myelencephalon
10. Isthmus
11. Myelencephalon
12. Metencephalon
13. Mesencephalon
14. Notochord
15. Infundibulum
16. Rathke's pouch
17. Stomodeum
18. Diencephalon
19. Preoral gut
20. Oral membrane
21. Pharynx
22. Mandibular process
23. First aortic arch
24. Telencephalon
25. Bulbus cordis
26. Ventricle
27. Aortic sac
28. Sensory retina
29. Pigmented retina
30. Lens vesicle
31. Opticoel (continuous with cavity of diencephalon)
32. Optic fissure
33. Head mesenchyme

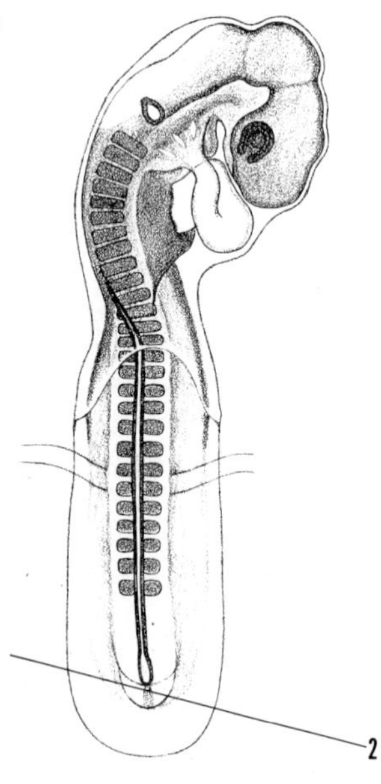

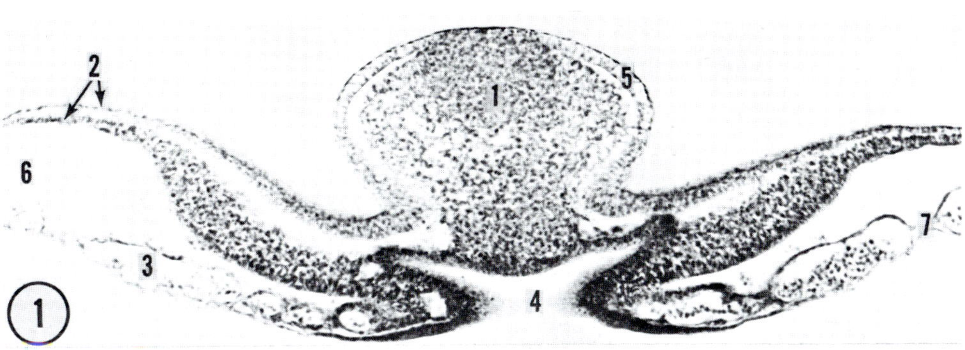
1
2
5
6
7
3
4

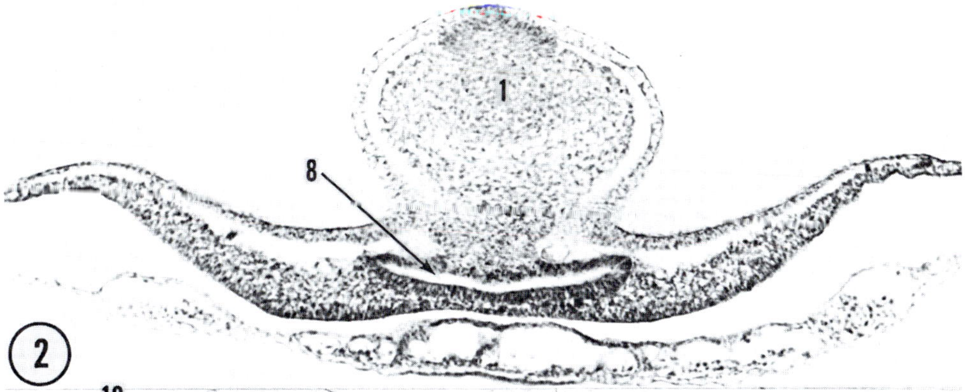
2
1
8

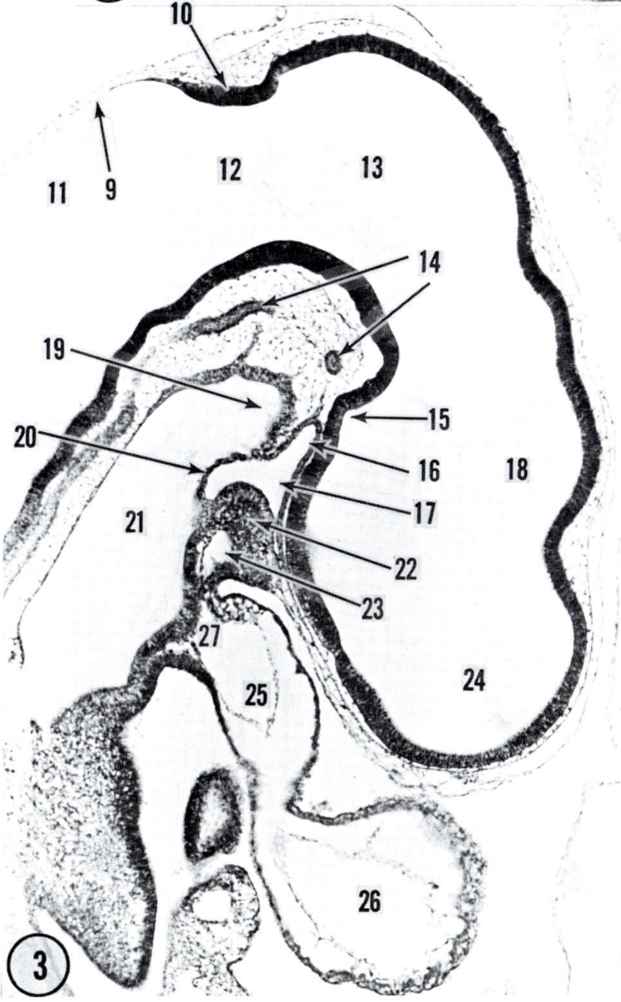
3
10
9
11
12
13
14
19
15
20
16
18
17
21
22
23
27
24
25
26

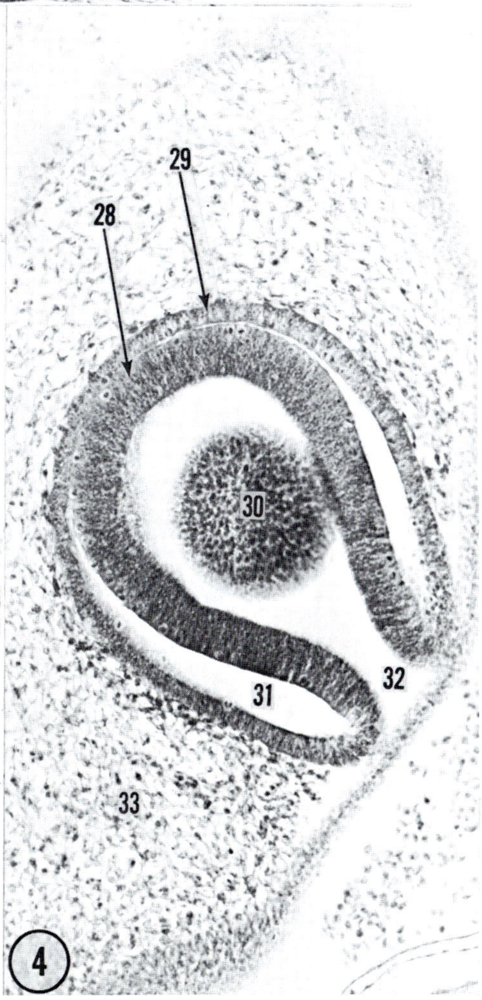
4
29
28
30
31
32
33

PLATE 14

72-HOUR CHICK EMBRYO WHOLE MOUNTS

1. Ventricle
2. Cerebral hemisphere of telencephalon
3. Pineal gland
4. Diencephalon
5. Mesencephalon
6. Isthmus
7. Metencephalon
8. Myelencephalon
9. Endolymphatic duct
10. Auditory vesicle
11. Wing bud
12. Cervical flexure
13. Vitelline artery
14. Caudal intestinal portal
15. Boundary of amniotic folds
16. Nasal pit
17. Semilunar ganglion
18. Acoustico-facialis ganglion
19. Cranial intestinal portal
20. Tail flexure
21. First branchial groove
22. Roof plate of myelencephalon
23. Third branchial arch
24. Atrium
25. Leg bud
26. Allantois
27. Tail
28. Maxillary process
29. Optic cup
30. Mandibular process
31. Lens vesicle
32. Second branchial arch
33. Second branchial groove
34. Third branchial groove
35. Bulbus cordis
36. Sinus venosus

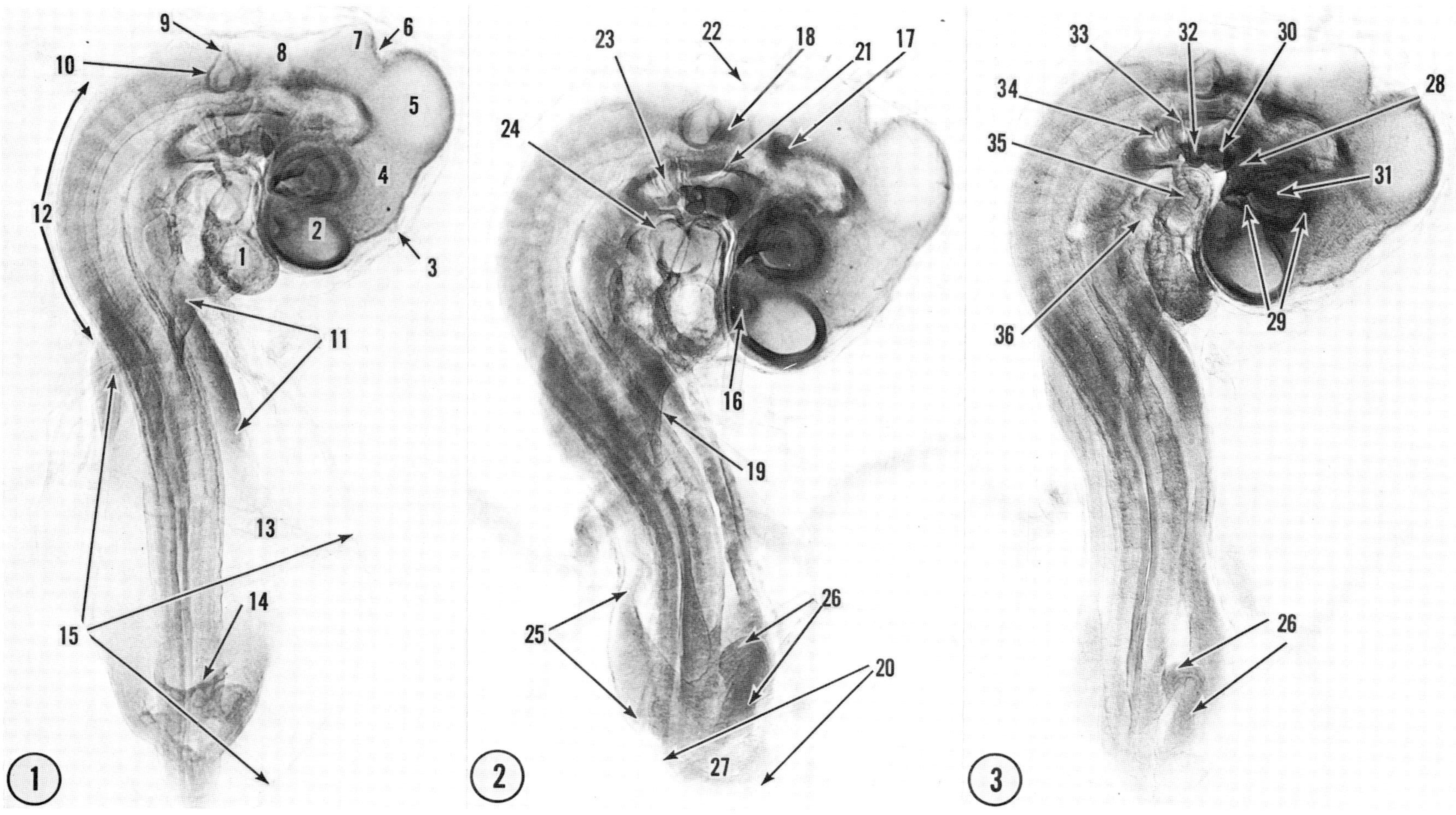
1
2
3
4
5
6
7
8
9
10
11
12
13
14
15
16
17
18
19
20
21
22
23
24
25
26
27
26
28
29
30
31
32
33
34
35
36

PLATE 15

72-HOUR CHICK EMBRYO WHOLE MOUNT

(enlargement of head region)

1. Second branchial arch
2. Mandibular process
3. Pharynx
4. Semilunar ganglion
5. Rathke's pouch
6. Infundibulum
7. Stomodeum
8. Lens vesicle
9. Diencephalon
10. Cerebral hemisphere of telencephalon

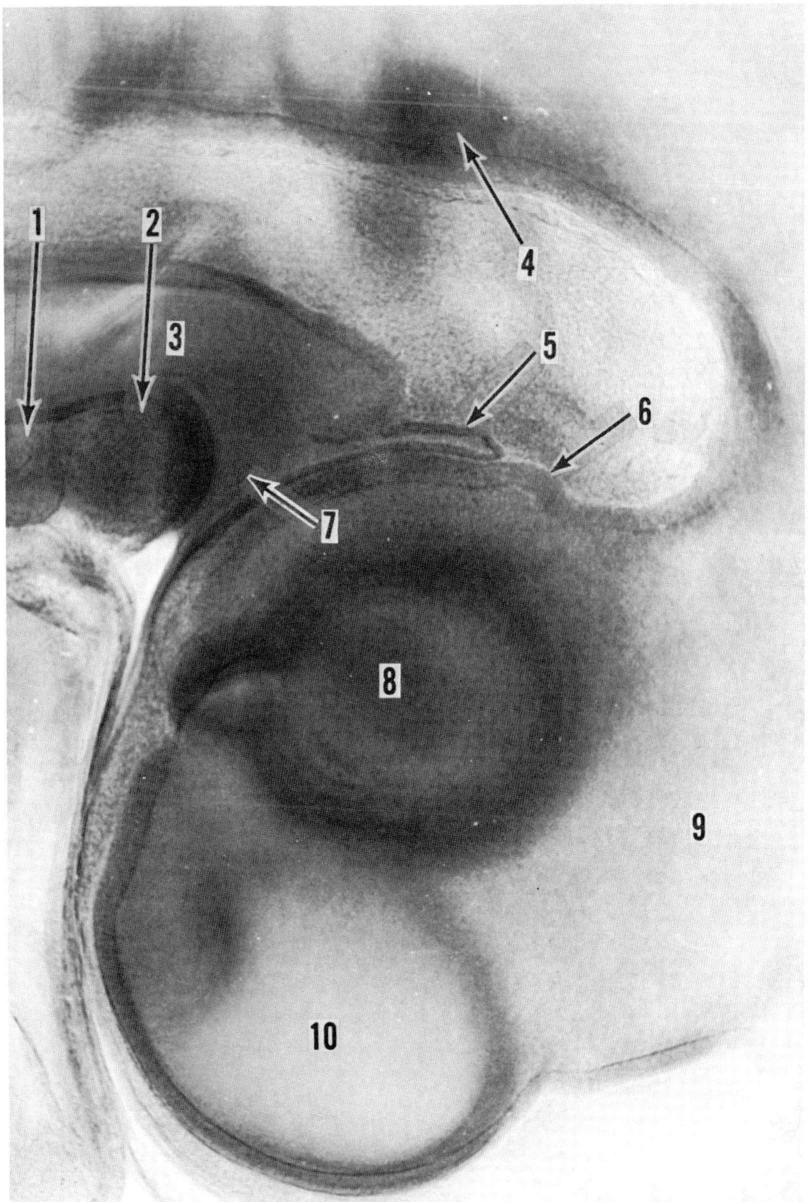
1
2
3
4
5
6
7
8
9
10

PLATE 16

72-HOUR INJECTED CHICK EMBRYO WHOLE MOUNT AND MOST ANTERIOR TRANSVERSE SECTION ILLUSTRATED

FIGURES 1, 2 Injected embryo whole mounts; Figure 2 is an enlargement of the heart and aortic arch region of the embryo illustrated in Figure 1

FIGURE 3 Most anterior transverse section illustrated

1. First aortic arch
2. Second aortic arch
3. Third aortic arch
4. Internal carotid artery
5. Dorsal aorta
6. Descending aorta
7. Vitelline artery
8. Fourth aortic arch
9. Aortic sac
10. Bulbus cordis
11. Ventricle
12. Yolk sac
13. Somite #1
14. Superior ganglion
15. Myelencephalon
16. Auditory vesicle
17. Acoustico-facialis ganglion
18. Precardinal vein
19. Neuromeres
20. Amniotic cavity
21. Semilunar ganglion
22. Metencephalon
23. Mesencephalon

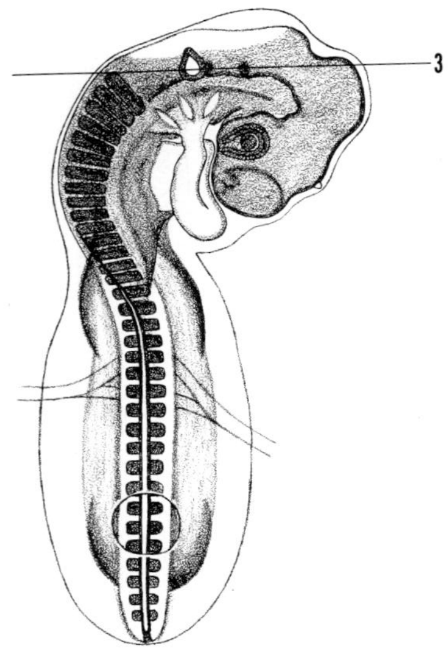

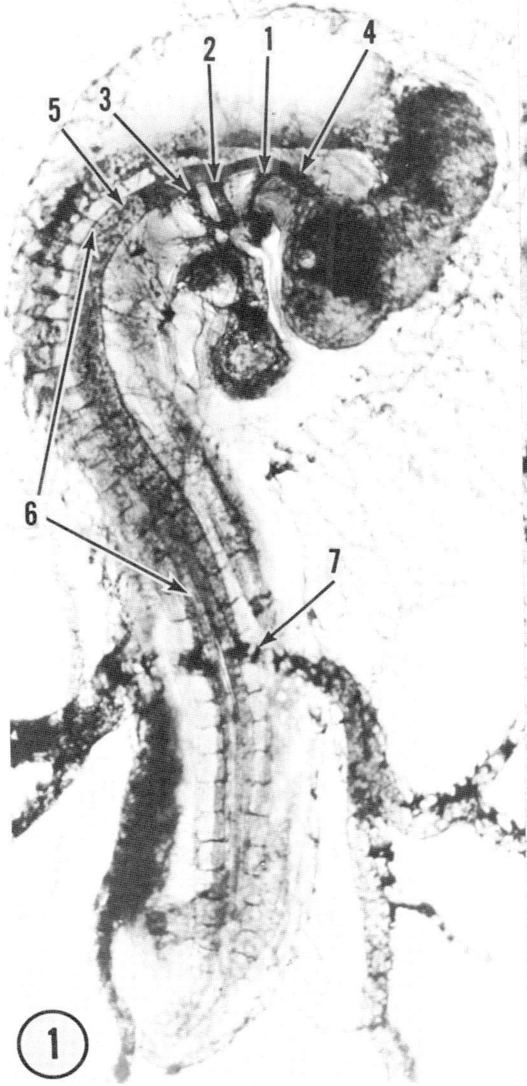

(1)

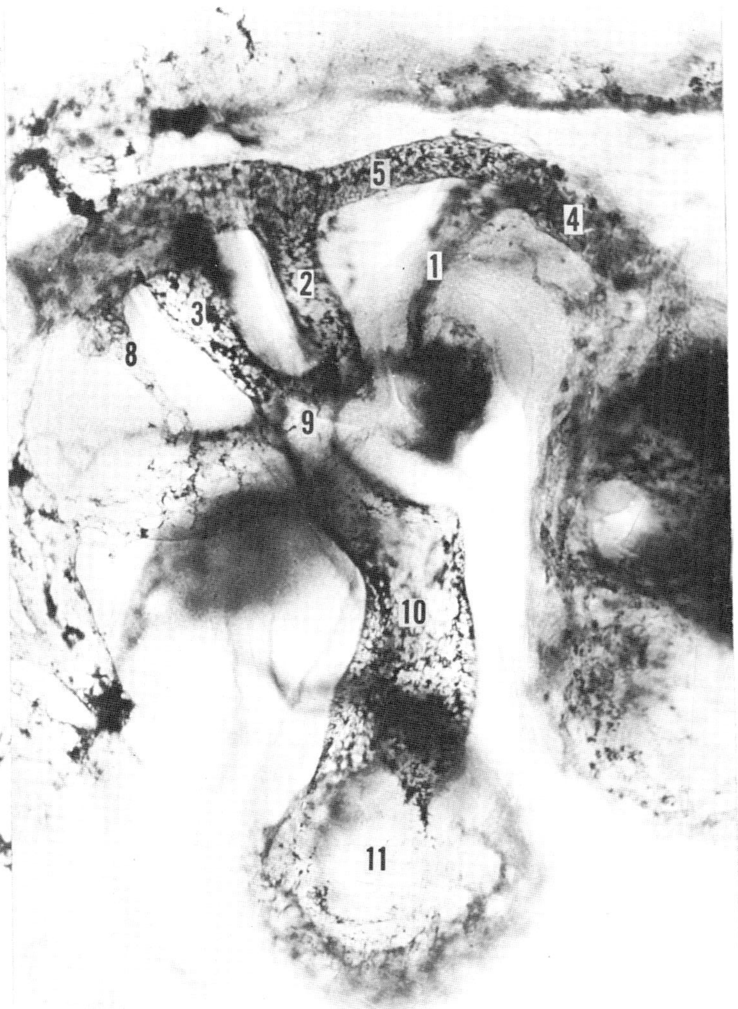

(2)

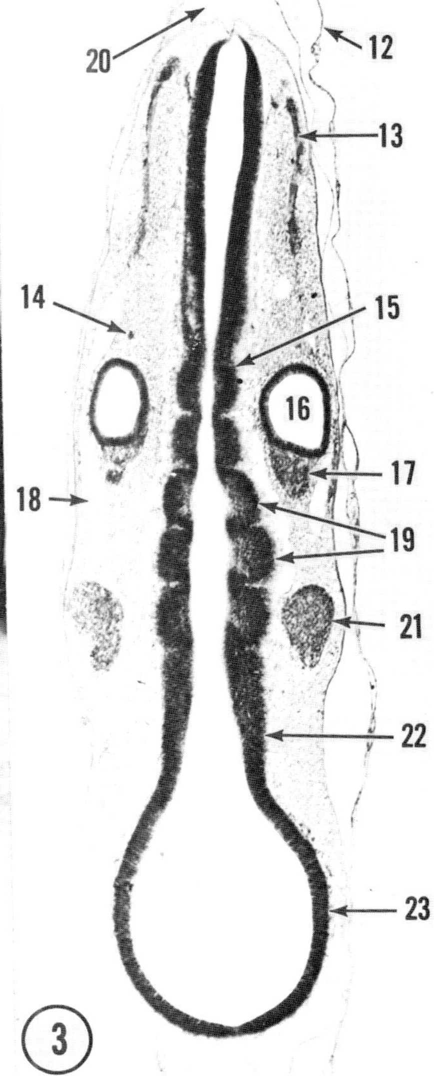

(3)

PLATE 17

72-HOUR CHICK EMBRYO SERIAL TRANSVERSE SECTIONS

1. Amnion
2. Yolk sac
3. Amniotic cavity
4. Somite
5. Notochord
6. Precardinal vein
7. Glossopharyngeal nerve
8. Epibranchial placode (contributes cells to facialis part of acoustico-facialis ganglion)
9. Facial nerve
10. Trigeminal nerve – maxillary branch
11. Trigeminal nerve – mandibular branch
12. Trigeminal nerve – ophthalmic branch
13. Mesencephalon
14. Extra-embryonic coelom
15. Skin ectoderm
16. Dorsal aorta
17. Second pharyngeal pouch
18. First pharyngeal pouch
19. First branchial cleft
20. Internal carotid artery
21. Third branchial arch
22. Jugular ganglion
23. Third aortic arch
24. Epibranchial placode (contributes cells to petrosal ganglion)
25. Second closing plate
26. Second aortic arch
27. Second branchial arch
28. Pharynx
29. First branchial groove
30. First closing plate
31. First aortic arch

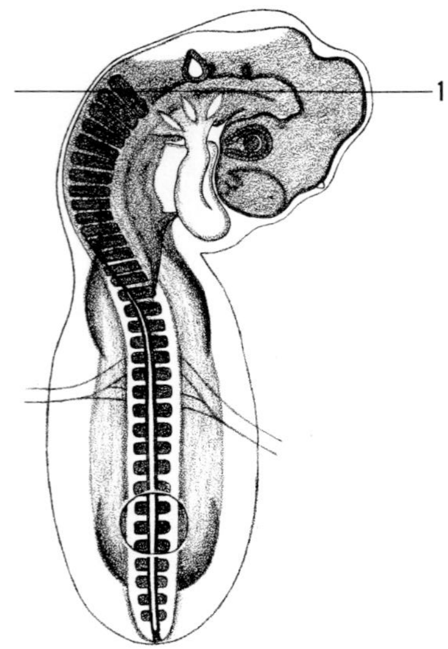

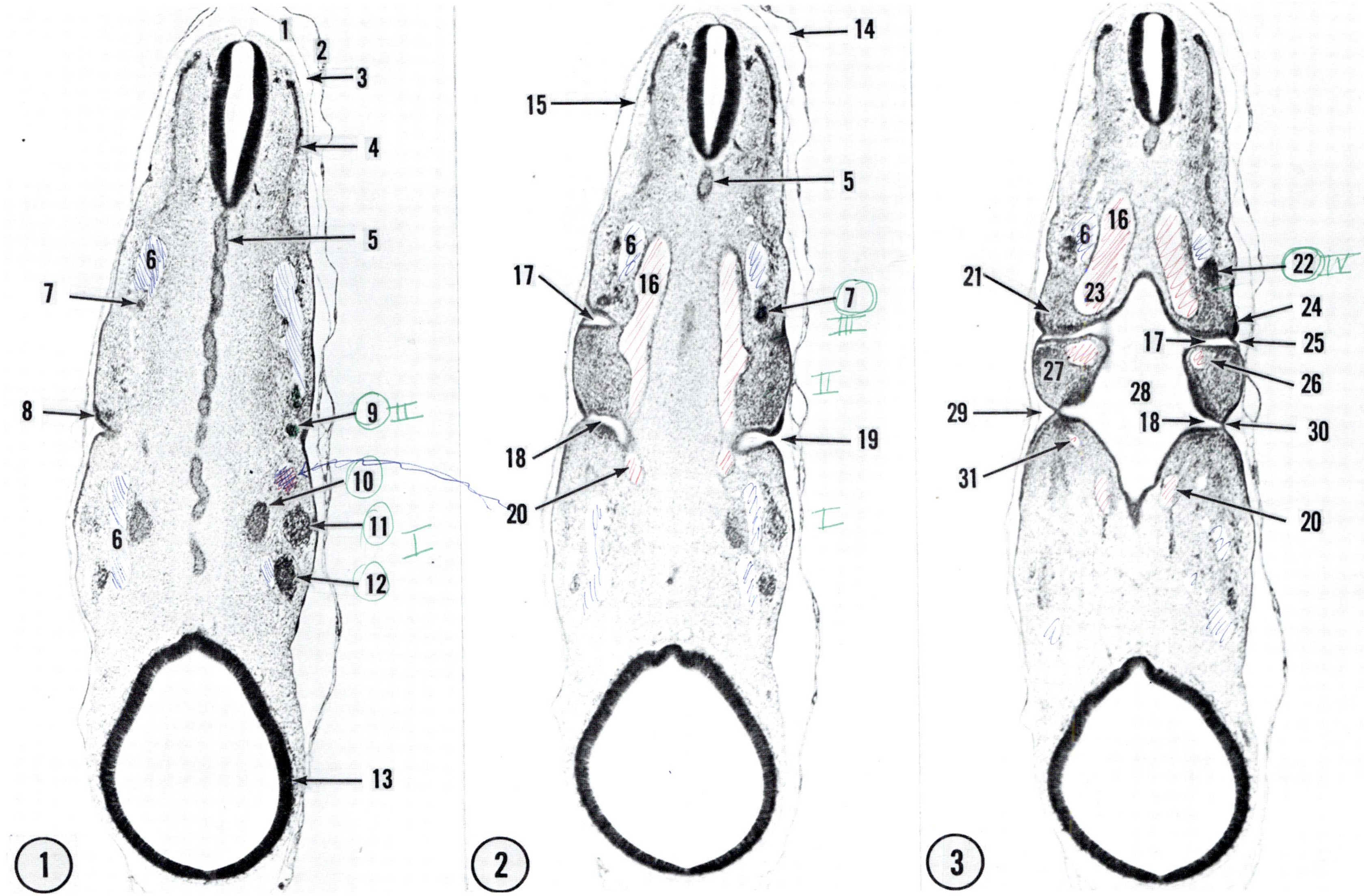
1
2
3
4
5
6
7
8
9
10
11
12
13
14
15
16
17
18
19
20
21
22
23
24
25
26
27
28
29
30
31

PLATE 18

72-HOUR CHICK EMBRYO SERIAL TRANSVERSE SECTIONS

1. Descending aorta
2. Dorsal aorta
3. Third pharyngeal pouch
4. Third aortic arch
5. Pharynx
6. Second pharyngeal pouch
7. Second branchial groove
8. Second aortic arch
9. First pharyngeal pouch
10. First aortic arch
11. Preoral gut
12. First branchial arch
13. Precardinal vein
14. Mesencephalon
15. Intersegmental artery
16. Fourth aortic arch
17. Third closing plate
18. Third branchial arch
19. Thyroid rudiment
20. Second branchial arch
21. Fourth branchial arch
22. Mandibular process
23. Maxillary process
24. Stomodeum
25. Rathke's pouch
26. Infundibulum
27. Spinal ganglion
28. Spinal cord
29. Notochord
30. Fourth closing plate
31. Fourth pharyngeal pouch
32. Vitelline blood vessel
33. Chorion
34. Amnion
35. Diencephalon

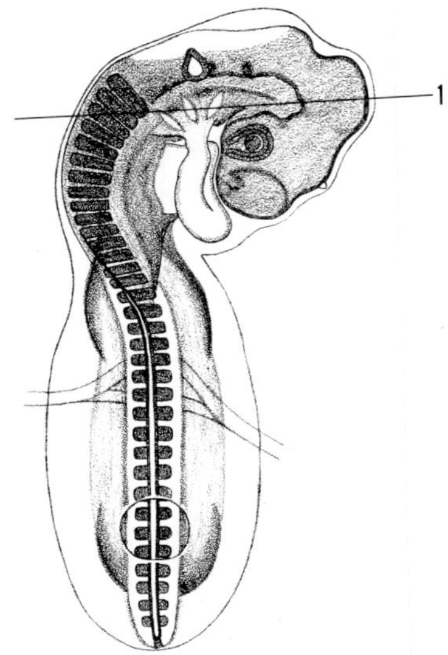

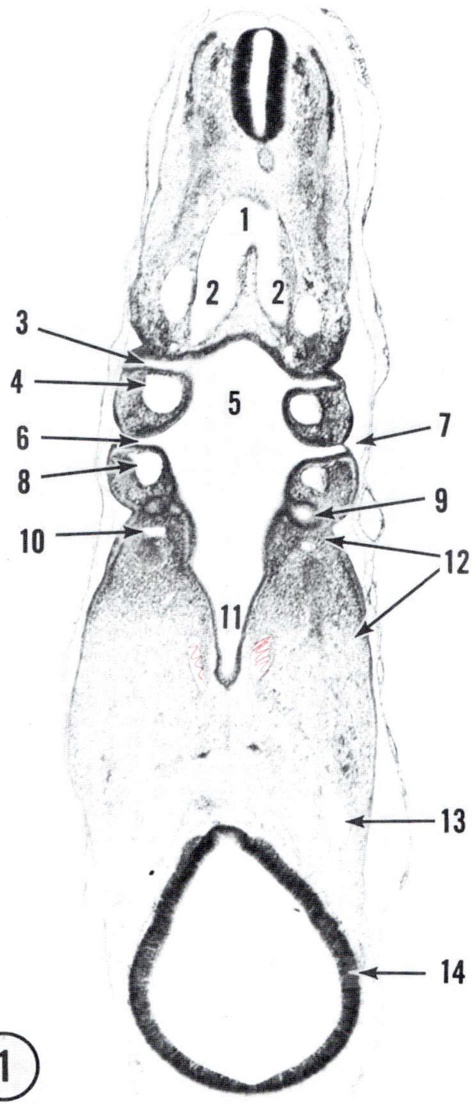
1
2
2
3
4
5
6
7
8
9
10
11
12
13
14

1

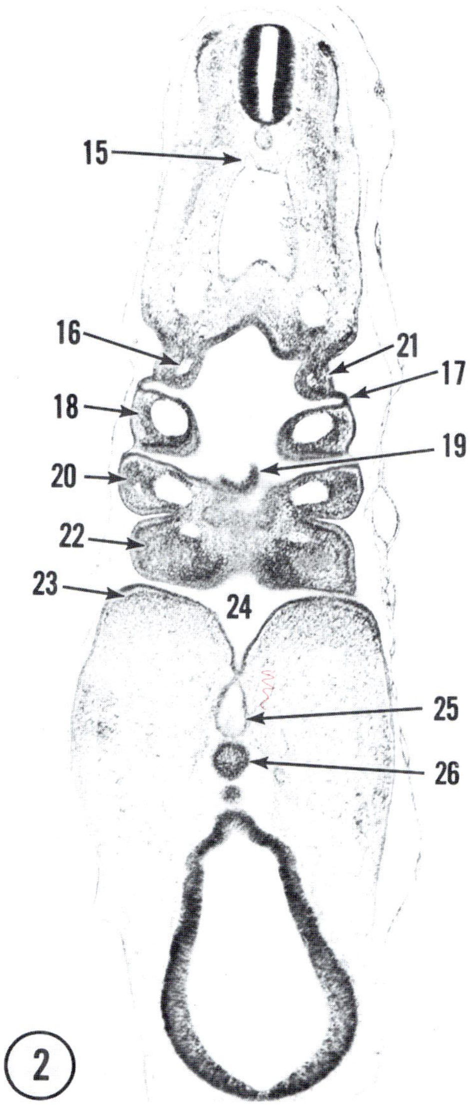
15
16
21
17
18
19
20
22
23
24
25
26

2

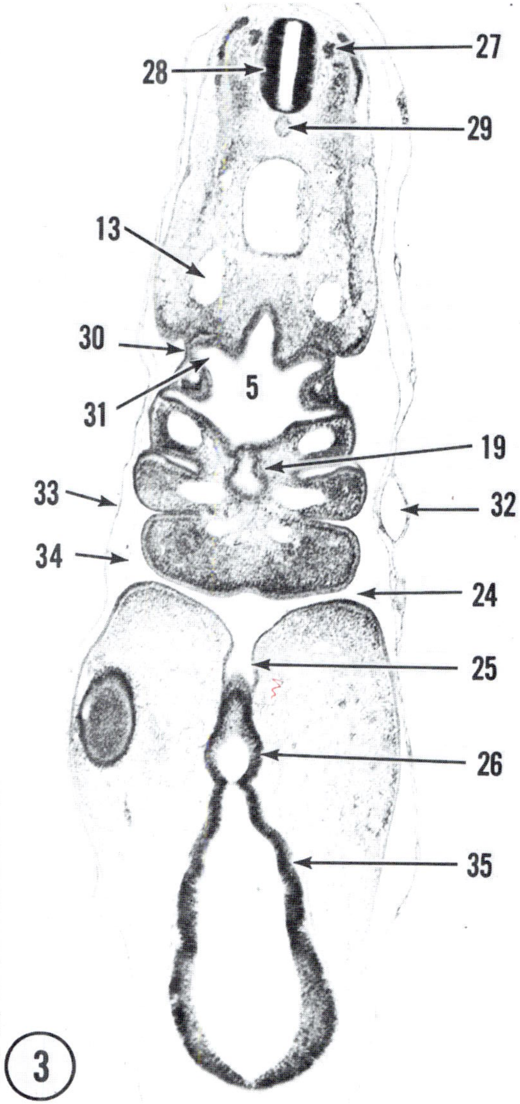
27
28
29
13
30
5
31
19
33
32
34
24
25
26
35

3

PLATE 19

72-HOUR CHICK EMBRYO SERIAL TRANSVERSE SECTIONS

1. Spinal cord
2. Postcardinal vein
3. Descending aorta
4. Common cardinal vein
5. Pharynx
6. Laryngo-tracheal groove
7. Aortic sac
8. Future corneal epithelium
9. Lens epithelium
10. Lens fibers
11. Ventral lip of optic cup
12. Dorsal lip of optic cup
13. Diencephalon
14. Spinal ganglion
15. Extra-embryonic coelom
16. Atrium
17. Optic fissure (note lack of ventral lip of optic cup)
18. Pigmented retina
19. Sensory retina
20. Head mesenchyme
21. Lung bud
22. Sinus venosus
23. Pericardial cavity
24. Epi-myocardium of bulbus cordis
25. Endocardium of bulbus cordis
26. Optic stalk

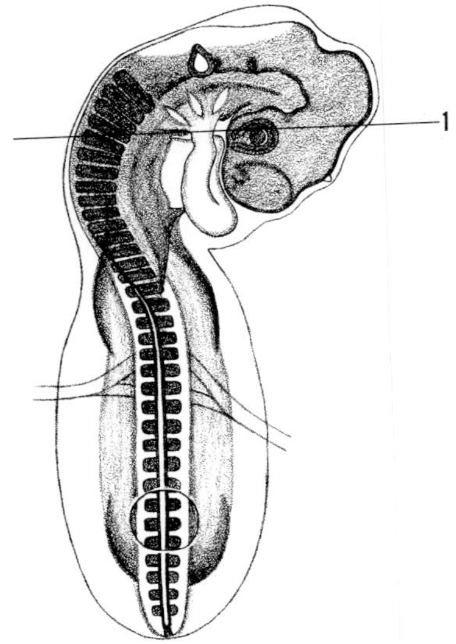

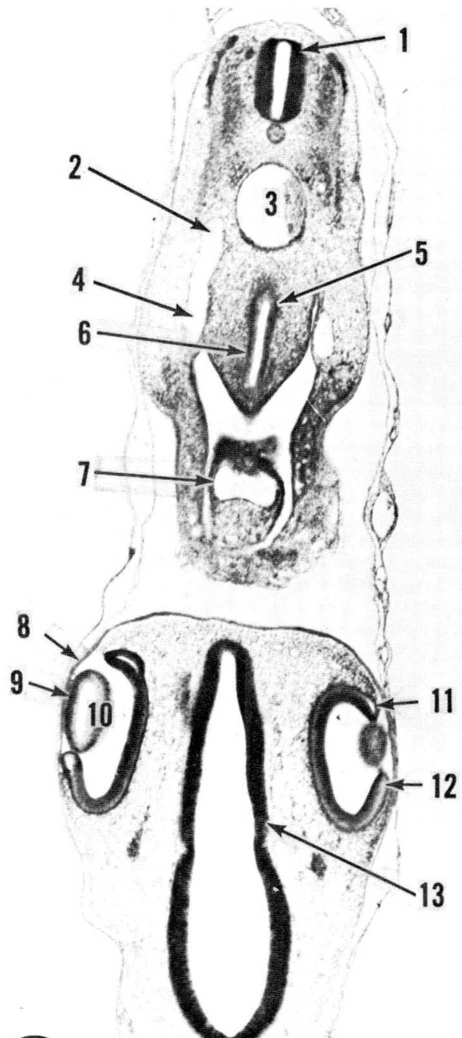
1
2
3
4
5
6
7
8
9
10
11
12
13

1

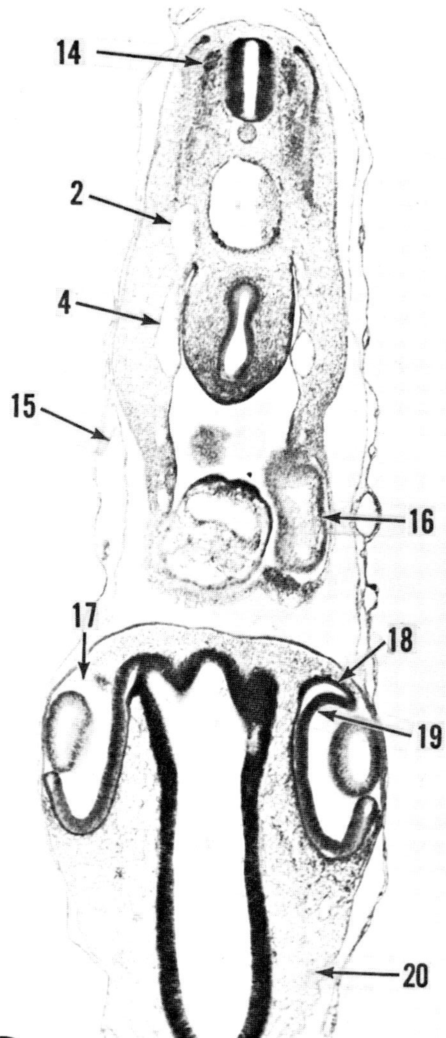
14
2
4
15
16
17
18
19
20

2

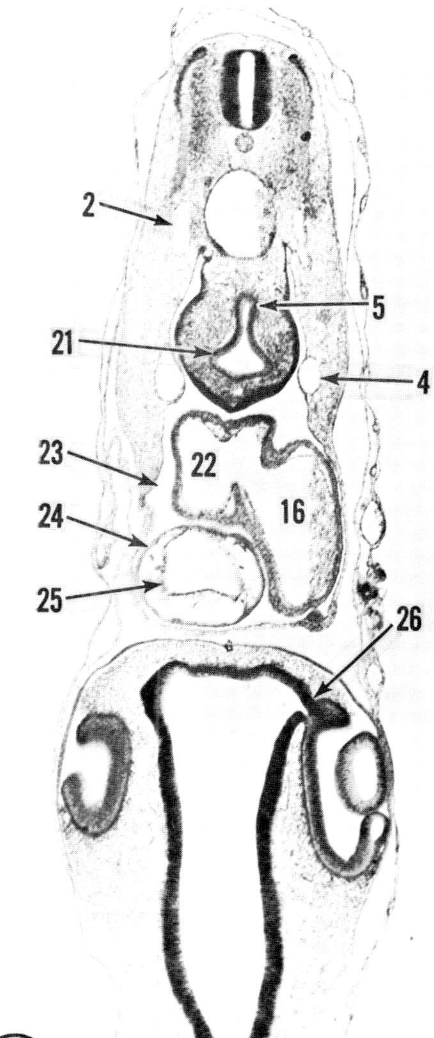
2
5
21
4
23
22
16
24
25
26

3

PLATE 20

72-HOUR CHICK EMBRYO SERIAL TRANSVERSE SECTIONS

1. Descending aorta
2. Postcardinal vein
3. Esophagus
4. Lung bud
5. Common cardinal vein
6. Sinus venosus
7. Atrium (continuous with ventricle below)
8. Bulbus cordis
9. Diencephalon
10. Mesoesophagus
11. Pleural cavity
12. Dorsal mesocardium
13. Somite
14. Notochord
15. Pericardial cavity
16. Cerebral hemisphere of telencephalon
17. Mesonephric duct (degenerating portion)
18. Dorsal mesogaster
19. Stomach
20. Gastro-hepatic ligament
21. Cranial liver rudiment
22. Ductus venosus
23. Endocardium of ventricle
24. Epi-myocardium of ventricle
25. Nasal pit
26. Pineal gland

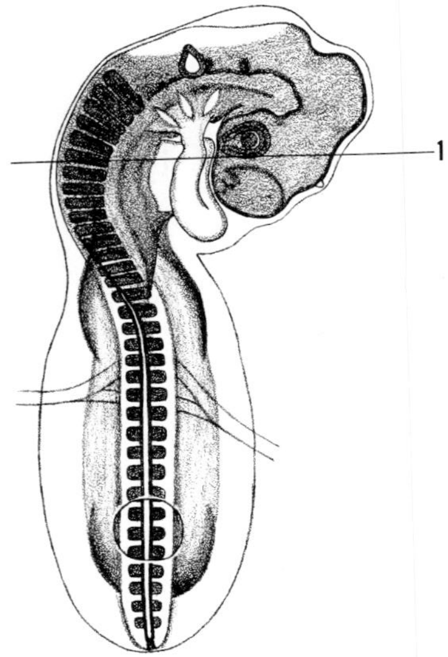

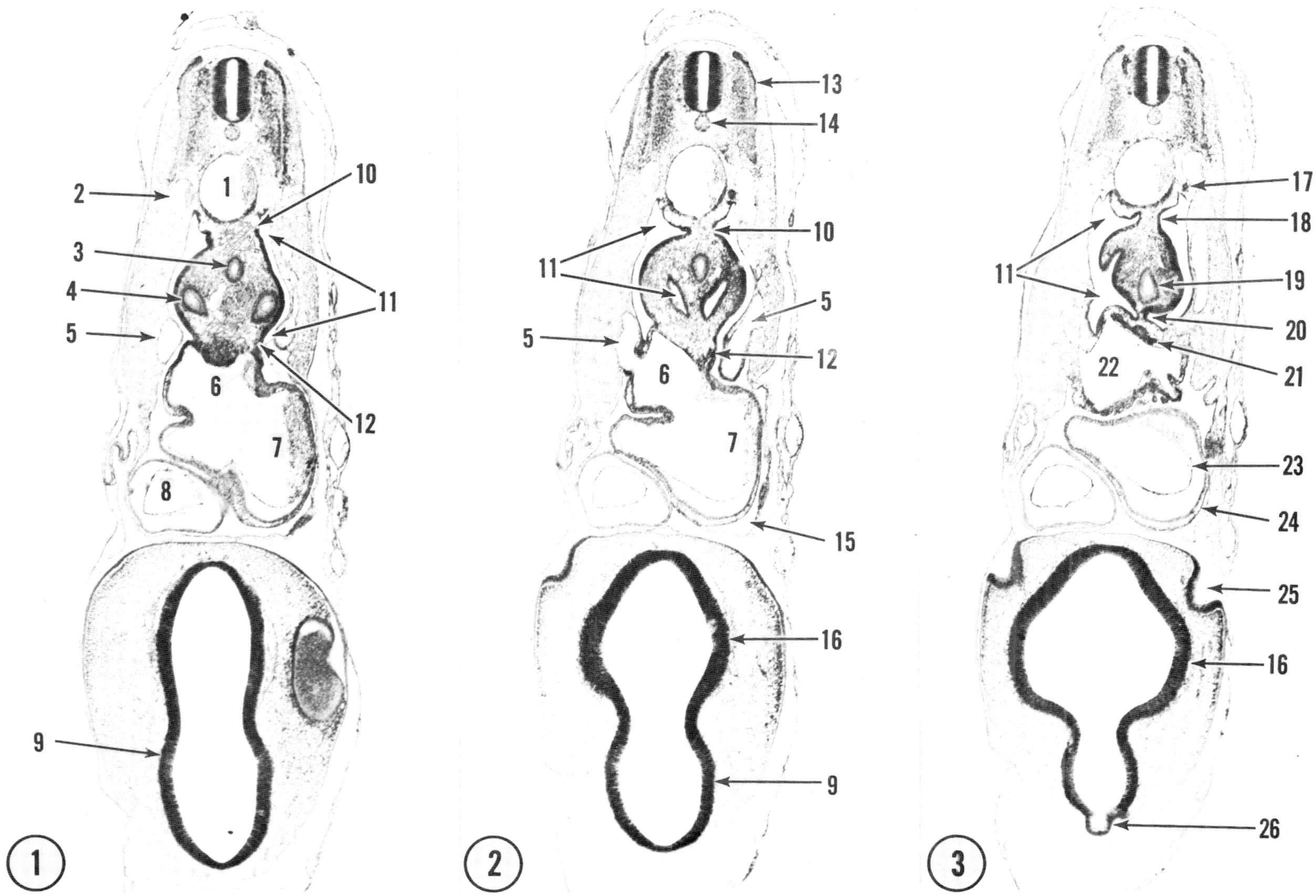
1
2
3
4
5
6
7
8
9
10
11
12
13
14
15
16
17
18
19
20
21
22
23
24
25
26

PLATE 21

72-HOUR CHICK EMBRYO SERIAL TRANSVERSE SECTIONS

1. Spinal ganglion
2. Postcardinal vein
3. Mesonephric duct (degenerating portion)
4. Mesoduodenum
5. Duodenum
6. Cranial liver rudiment
7. Peritoneal cavity
8. Ductus venosus
9. Caudal liver rudiment
10. Bulbus cordis
11. Ventricle
12. Cerebral hemisphere of telencephalon
13. Duodeno-hepatic ligament
14. Spinal cord
15. Yolk sac
16. Amnion
17. Descending aorta
18. Dorsal pancreatic rudiment
19. Vitelline vein
20. Chorion
21. Dermatome
22. Myotome
23. Sclerotome
24. Intersegmental artery
25. Cranial intestinal portal

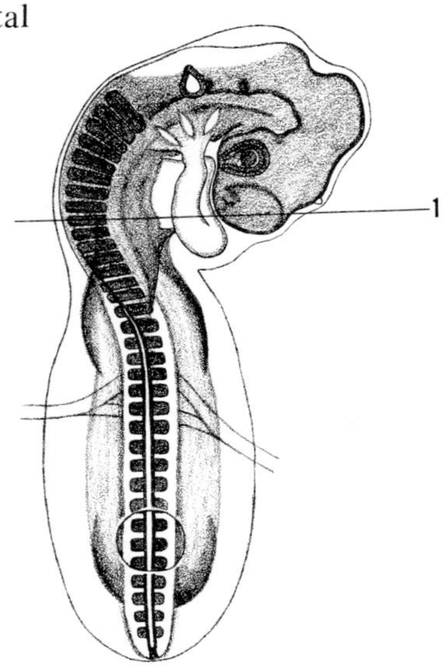

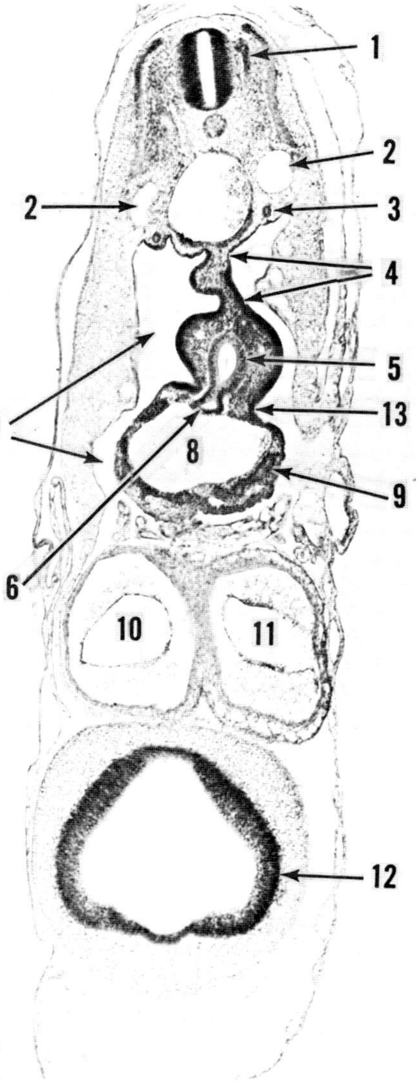
1
2
2
3
4
5
13
7
8
9
6
10
11
12

1

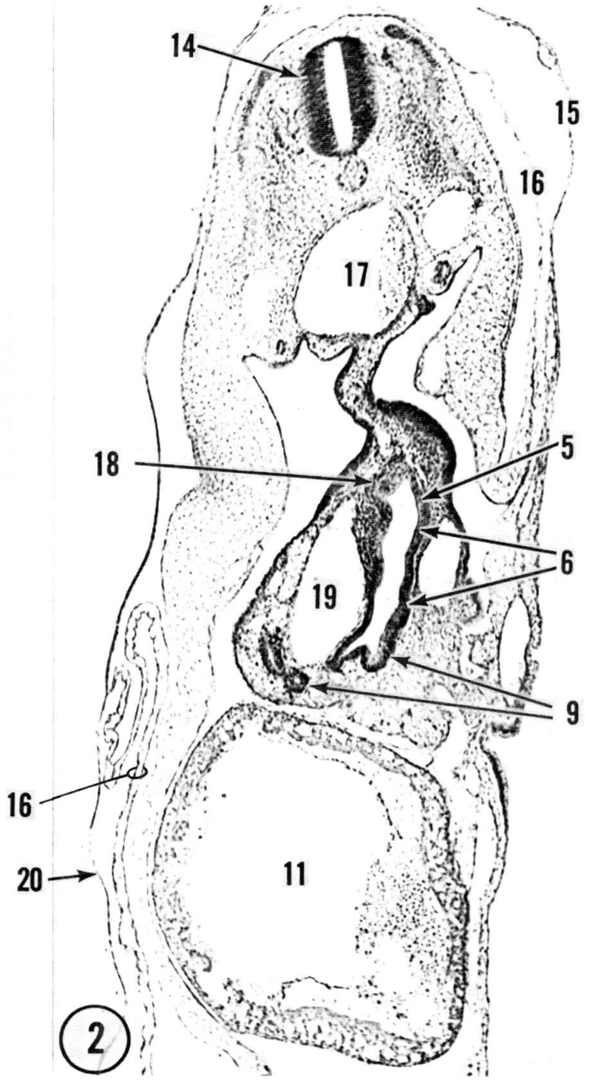
14
15
16
17
18
5
6
19
9
16
20
11

2

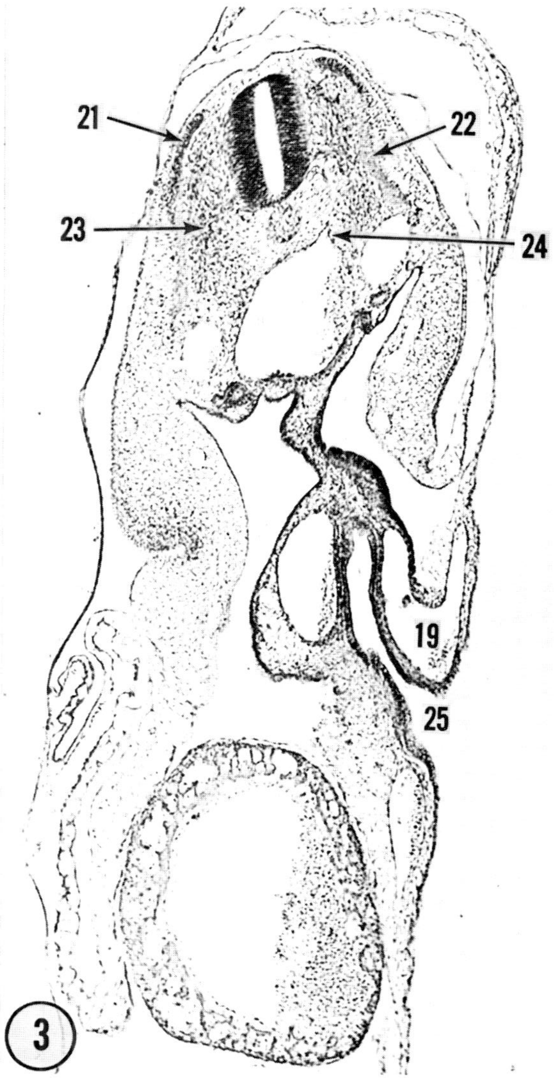
21
22
23
24
19
25

3

PLATE 22

72-HOUR CHICK EMBRYO SERIAL TRANSVERSE SECTIONS

1. Dermatome
2. Myotome
3. Sclerotome
4. Somatic mesoderm of *wing* bud
5. Mesonephric duct
6. Mesonephric tubule
7. Mesonephric tubule rudiment
8. Descending aorta
9. Apical ectodermal ridge of *wing* bud
10. Vitelline blood vessels
11. Postcardinal vein
12. Dorsal aortae
13. Somatic mesoderm of *leg* bud
14. Chorion
15. Amnion
16. Amniotic cavity
17. Extra-embryonic coelom
18. Yolk sac
19. Lateral amniotic fold
20. Peritoneal cavity
21. Hindgut
22. Cloaca
23. Apical ectodermal ridge of *leg* bud
24. Splanchnic mesoderm of allantois
25. Endoderm of allantois
26. Cloacal membrane
27. Subcaudal pocket (amniotic cavity)

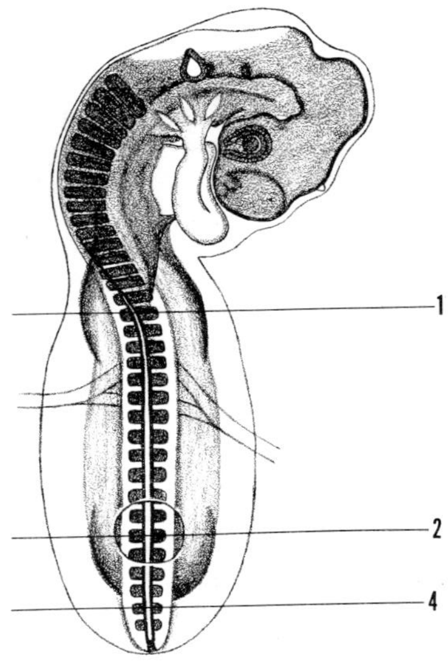

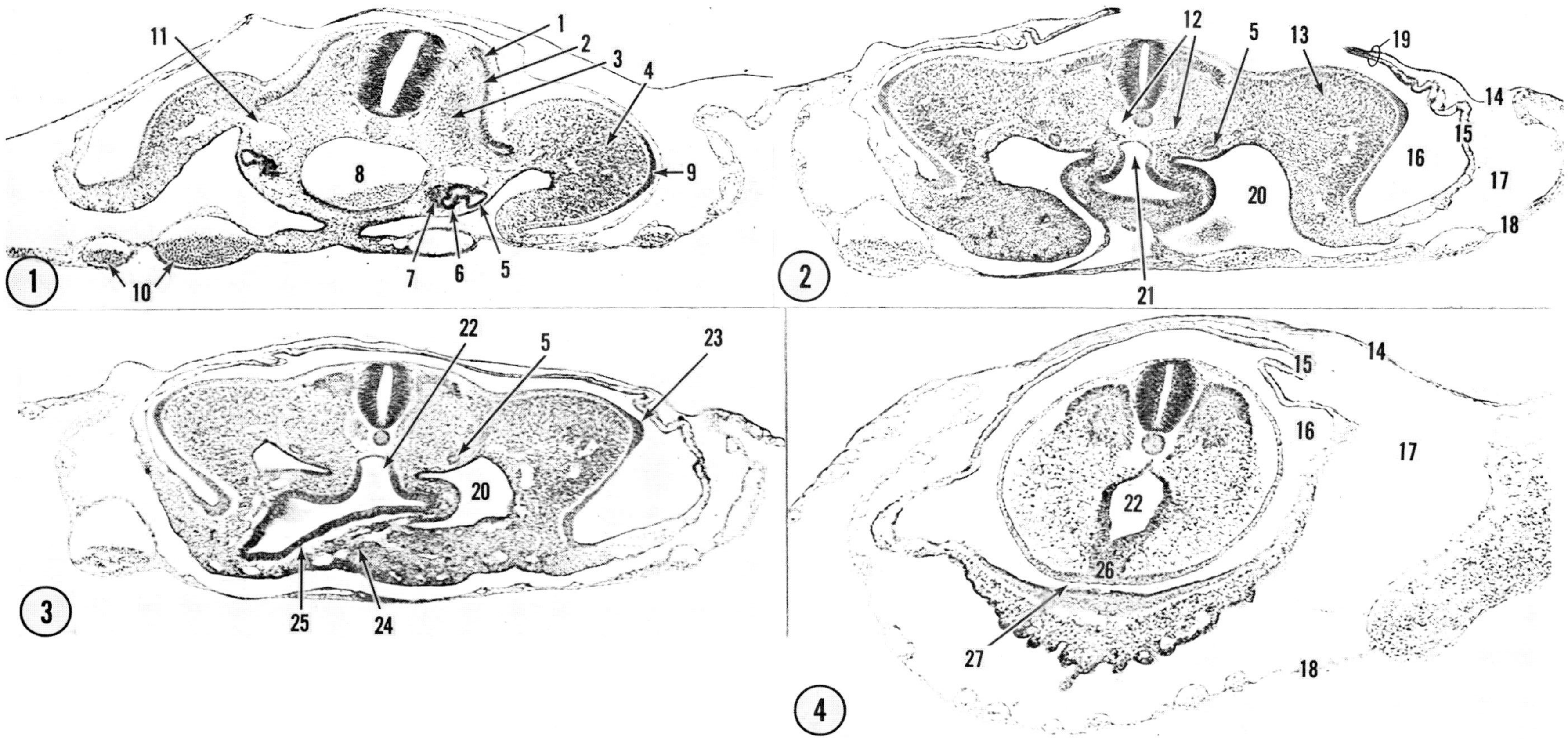
1
2
3
4
5
6
7
8
9
10
11
12
13
14
15
16
17
18
19
20
21
22
23
24
25
26
27
1
2
3
4

PLATE 23

72-HOUR CHICK EMBRYO MIDSAGITTAL SECTION

(enlargement of head region)

1. Isthmus
2. Myelencephalon
3. Notochord
4. Metencephalon
5. Mesencephalon
6. Preoral gut
7. Amniotic cavity
8. Rathke's pouch
9. Pharynx
10. Mouth opening
11. Stomodeum
12. Infundibulum
13. Diencephalon
14. Pineal gland
15. Telencephalon

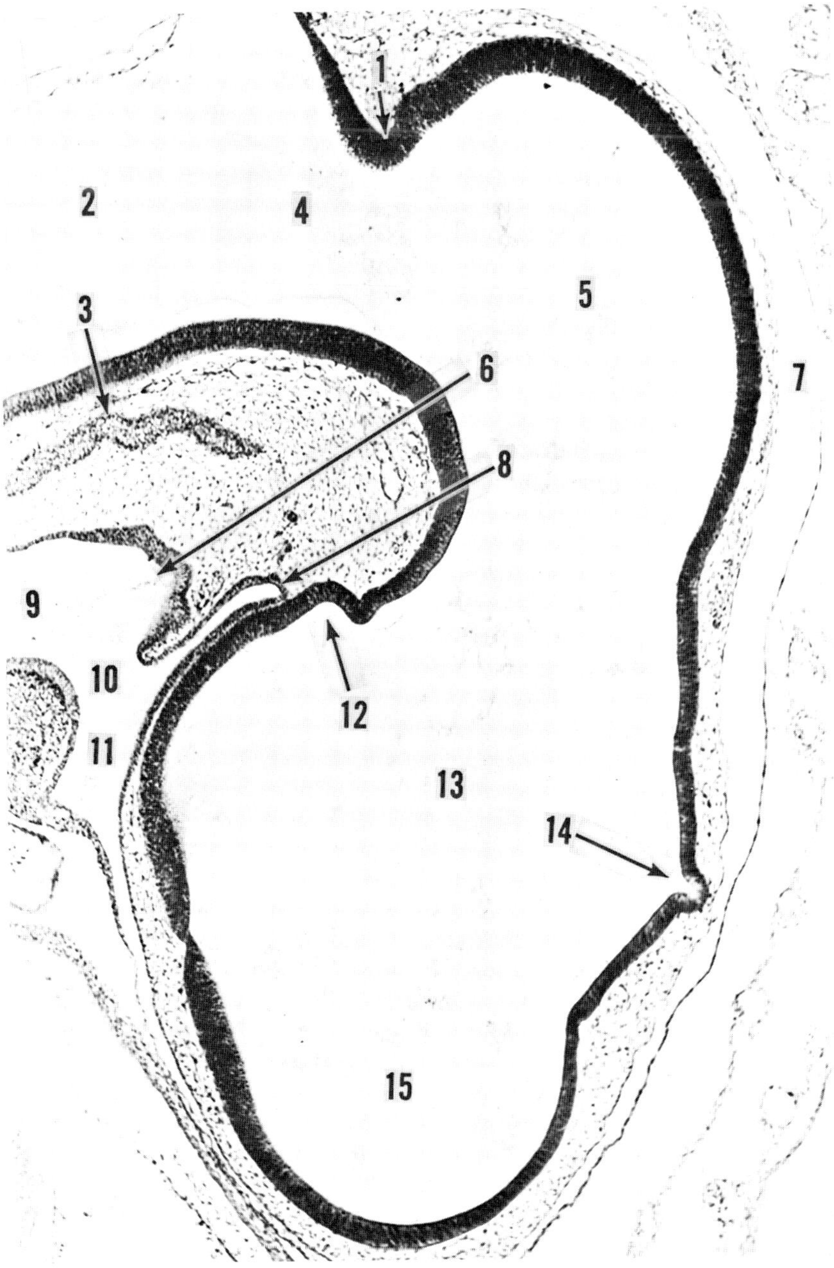
1
2
4
5
3
6
7
8
9
10
12
11
13
14
15

PLATE 24

10-MM PIG EMBRYO SERIAL TRANSVERSE SECTIONS

1. Amnion
2. Myelencephalon
3. Fourth ventricle
4. Metencephalon
5. Isthmus
6. Mesencephalon
7. Cerebral aqueduct
8. Ventricular zone
9. Intermediate zone
10. Marginal zone
11. Roots of the spinal accessory nerve
12. Pia mater
13. Endolymphatic duct
14. Spinal accessory nerve
15. Sensory root of the vagus nerve
16. Auditory vesicle (future utriculus)

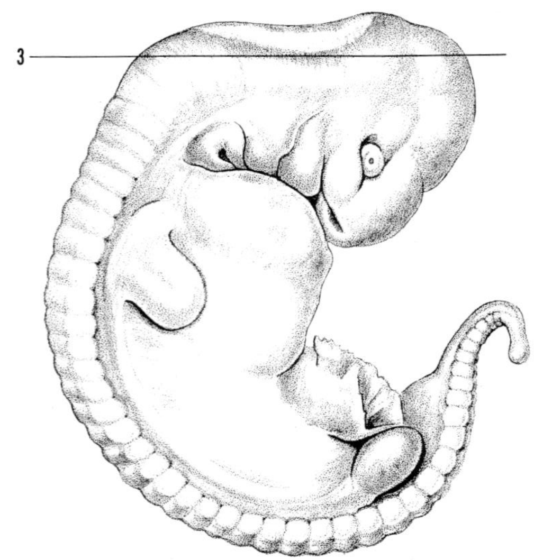

1

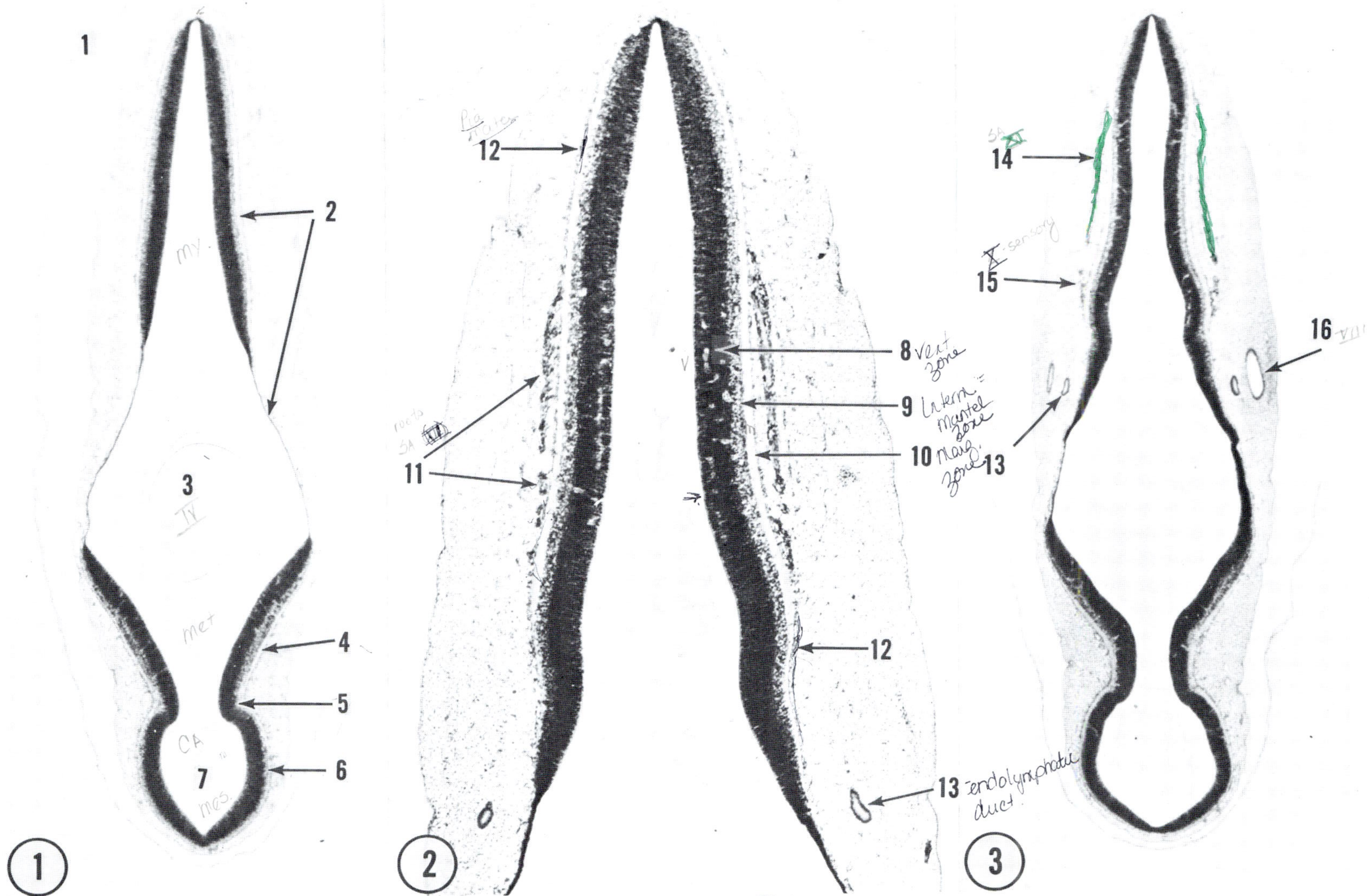

PLATE 25

10-MM PIG EMBRYO SERIAL TRANSVERSE SECTIONS

1. Spinal accessory nerve
2. Myelencephalon
3. Jugular ganglion
4. Sensory root of the glossopharyngeal nerve
5. Auditory vesicle (future utriculus)
6. Endolymphatic duct
7. Neuromeres
8. Metencephalon
9. Mesencephalon
10. Froriep's ganglion
11. Roots of the hypoglossal nerve
12. Superior ganglion
13. Amniotic cavity
14. Rudiment of the posterior semicircular canal
15. Rudiment of the anterior semicircular canal
16. Sensory root of the trigeminal nerve
17. Fourth ventricle
18. Cerebral aqueduct
19. Vagus nerve
20. Glossopharyngeal nerve
21. Precardinal vein
22. Auditory vesicle (future sacculus)
23. Acoustic ganglion
24. Geniculate ganglion
25. Semilunar ganglion
26. Rhombencephalon
27. Basilar artery
28. Oculomotor nerve
29. Posterior communicating artery

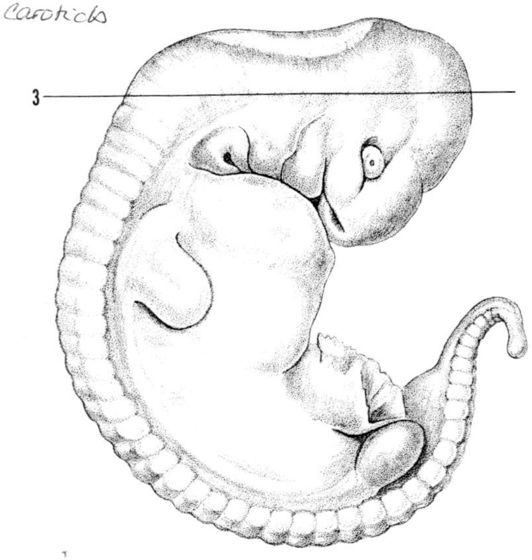

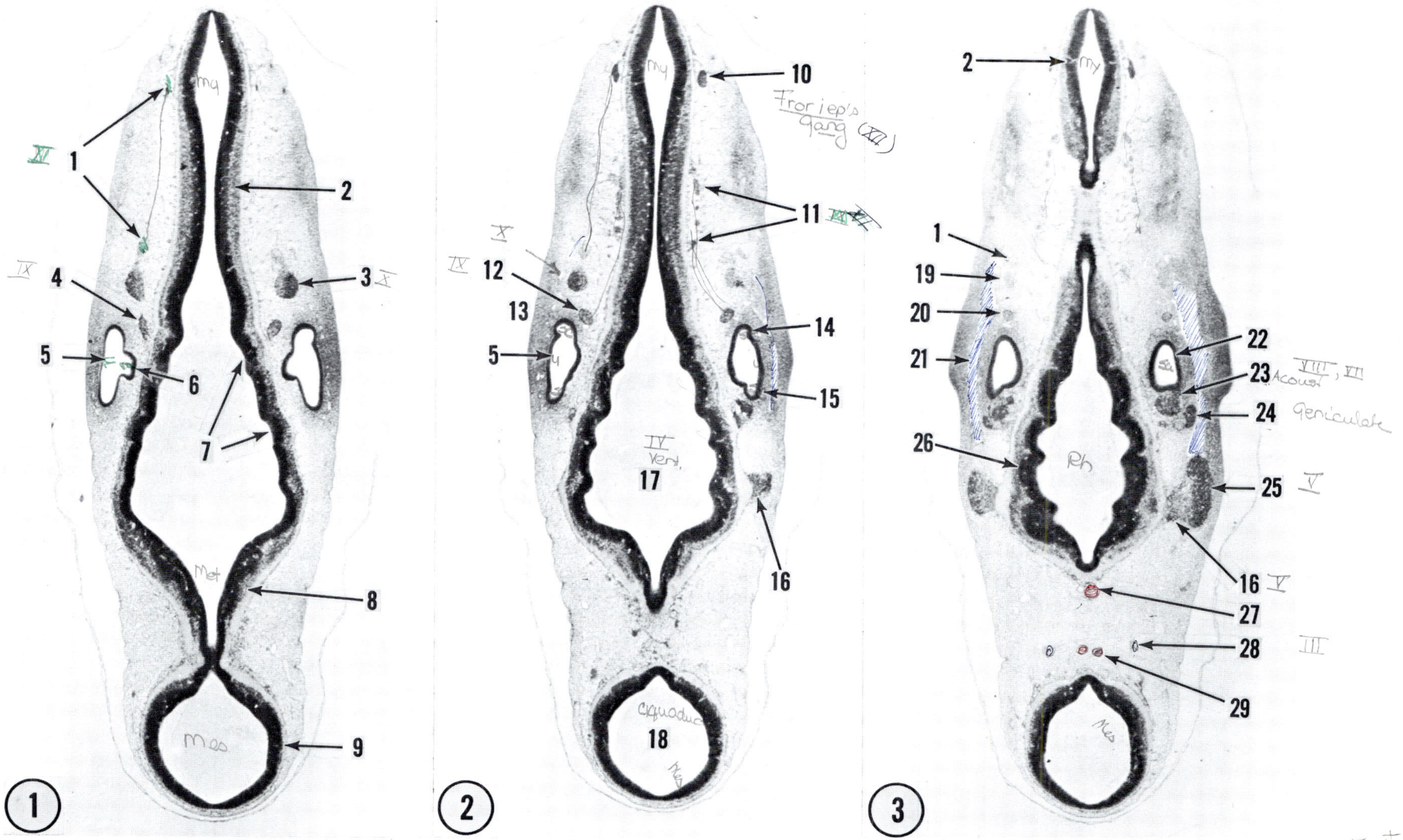
1
2
3
4
5
6
7
8
9
10
11
12
13
14
15
16
17
18
19
20
21
22
23
24
25
26
27
28
29

PLATE 26

10-MM PIG EMBRYO SERIAL TRANSVERSE SECTIONS

1. Roots of the hypoglossal nerve
2. Vertebral artery
3. Basilar artery
4. Precardinal vein
5. Acoustic ganglion
6. Sensory root of the facial nerve
7. Rhombencephalon
8. Semilunar ganglion
9. Oculomotor nerve
10. Posterior communicating artery
11. Mesencephalon
12. Spinal ganglion
13. Petrosal ganglion
14. Auditory vesicle (future cochlea)
15. Facial nerve
16. Abducens nerve
17. First closing plate
18. First pharyngeal pouch (future tympanic cavity)
19. Internal carotid artery (near level a, Fig. O)
20. First branchial groove
21. Infundibulum
22. Diencephalon
23. Third ventricle (major part)

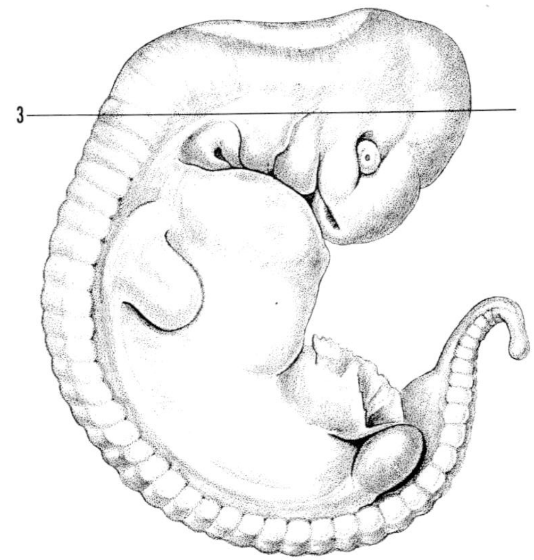

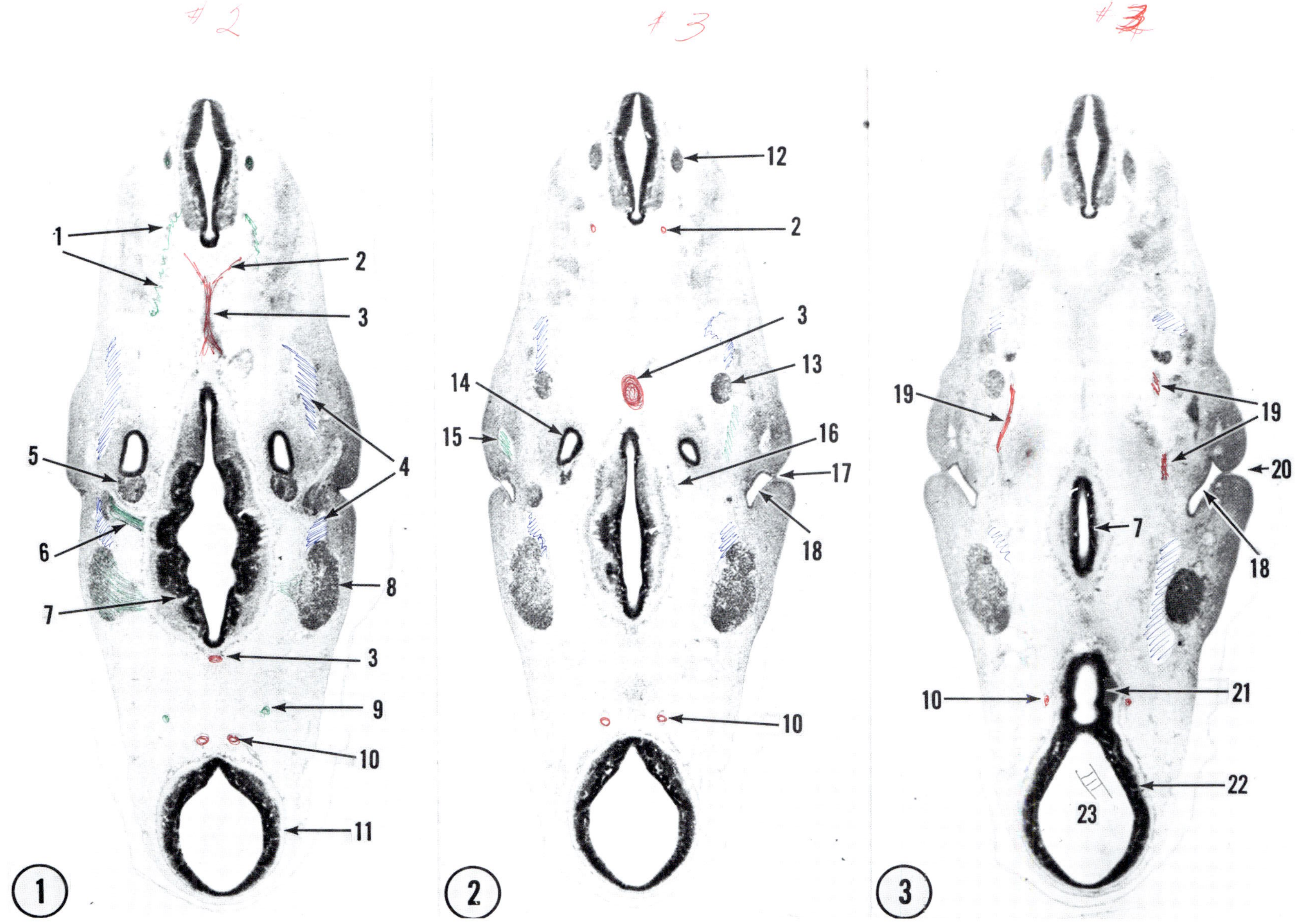

#2
#3
#3
1
2
3
4
5
6
7
8
3
9
10
11
12
2
3
13
14
15
16
17
18
10
19
19
20
7
18
10
21
22
23
III
1
2
3

PLATE 27

10-MM PIG EMBRYO SERIAL TRANSVERSE SECTIONS

1. Ductus caroticus
2. Third aortic arch
3. Second pharyngeal pouch
4. Pharynx
5. Second branchial arch
6. Facial nerve
7. First pharyngeal pouch (future eustachian tube, continuous below with the future tympanic cavity)
8. Basilar artery (near level f, Fig. O)
9. Internal carotid artery (near level b, Fig. O)
10. Mandibular branch of the trigeminal nerve
11. Semilunar ganglion
12. Posterior communicating artery
13. Glossopharyngeal nerve
14. Nodosum ganglion
15. Second branchial groove
16. Copula
17. Tuberculum impar
18. Eye muscle rudiments
19. Rathke's pouch
20. Infundibulum
21. Diencephalon
22. Precardinal vein
23. Spinal accessory nerve
24. Lateral lingual swelling
25. Internal carotid artery (near level c, Fig. O)
26. Anterior cerebral artery
27. Third ventricle (major part)

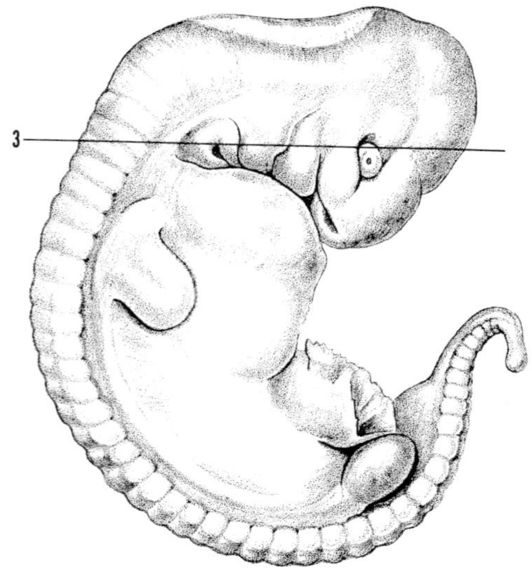

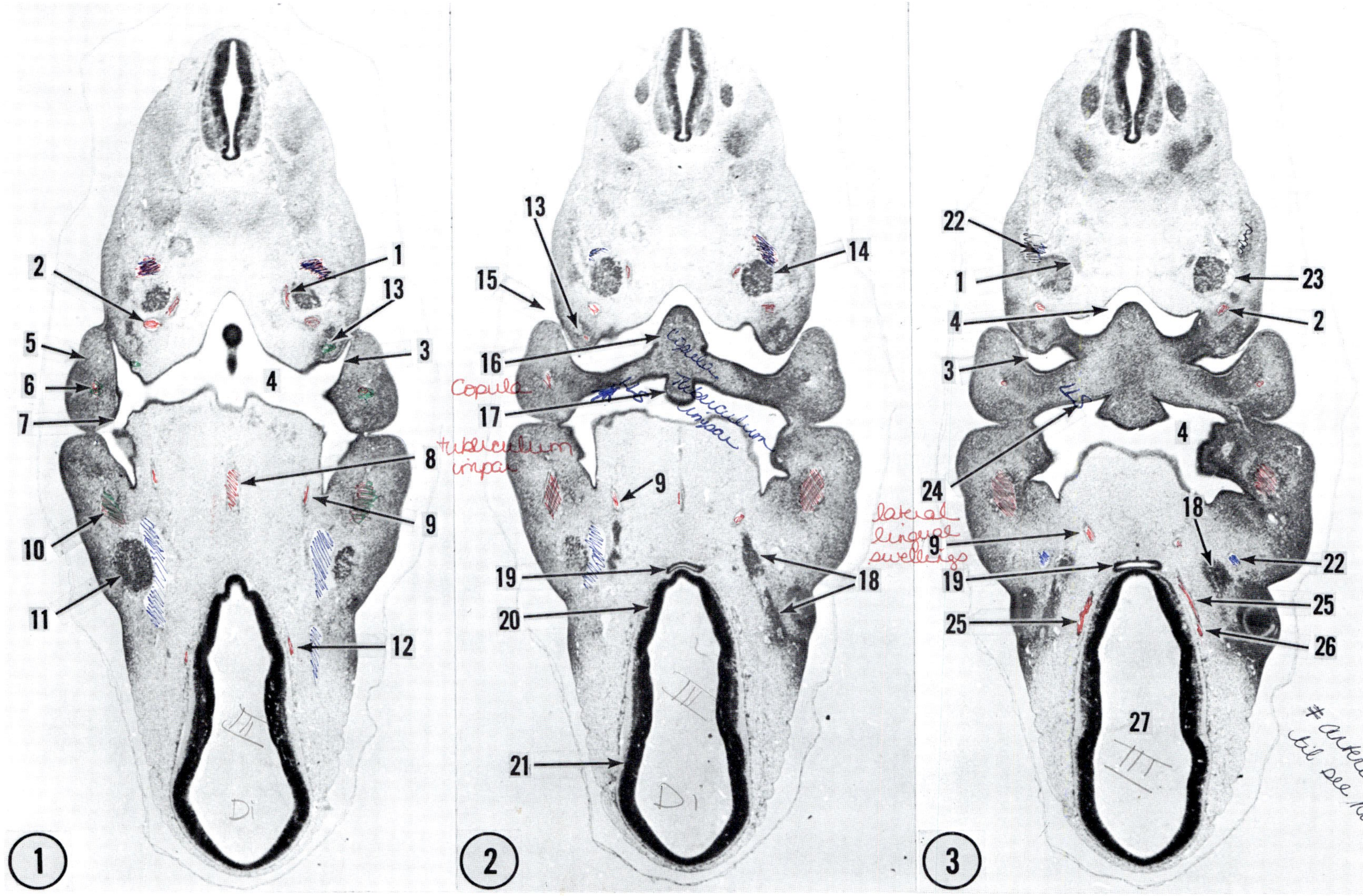
1
2
13
5
3
6
4
7
8
9
10
11
12
13
14
15
16
17
9
19
18
20
21
22
1
23
4
2
3
4
24
18
9
22
19
25
25
26
27
1
2
3

PLATE 28

10-MM PIG EMBRYO SERIAL TRANSVERSE SECTIONS

1. Epiglottis
2. Third pharyngeal pouch (inferior parathyroid rudiment)
3. Third aortic arch
4. Third branchial arch
5. Second branchial arch
6. Mandibular branch of the trigeminal nerve
7. Stomodeum
8. Sensory retina
9. Pigmented retina
10. Diencephalon
11. Third ventricle (major part)
12. Spinal cord
13. Spinal ganglion
14. Dorsal aorta
15. Fourth aortic arch
16. Cervical sinus
17. Third closing plate
18. Thyroid gland
19. Mandibular process
20. Maxillary process
21. Lens vesicle
22. Glottis
23. Arytenoid swelling
24. Precardinal vein
25. Fourth branchial arch
26. Third pharyngeal pouch (thymus rudiment)
27. Aortic trunk
28. Optic stalk
29. Optic fissure
30. Future corneal epithelium
31. Anterior cerebral artery
32. Pharynx

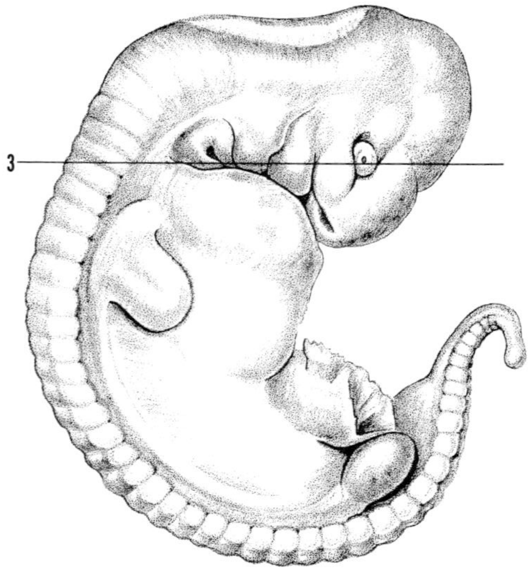

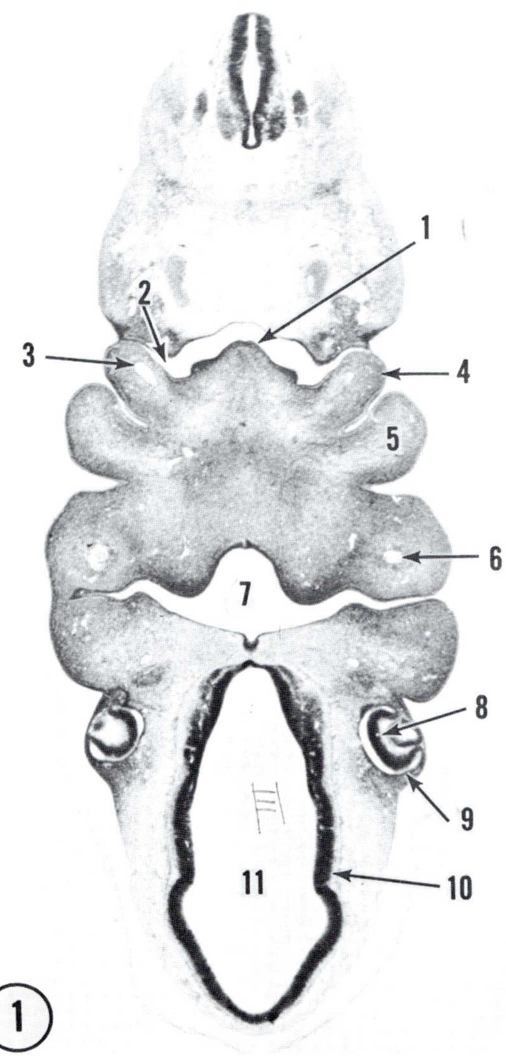

1
2
3
4
5
6
7
8
9
10
11

1

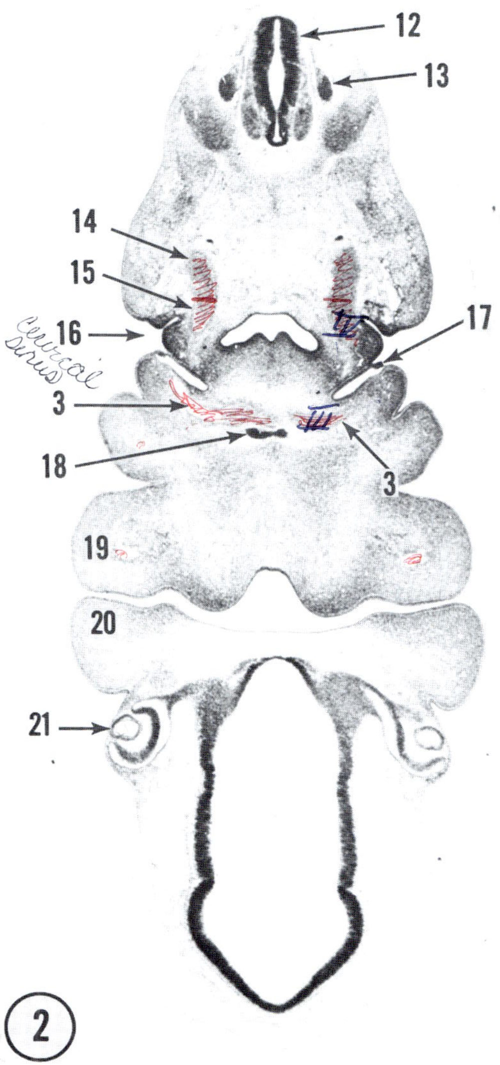

12
13
14
15
16
17
3
18
3
19
20
21

2

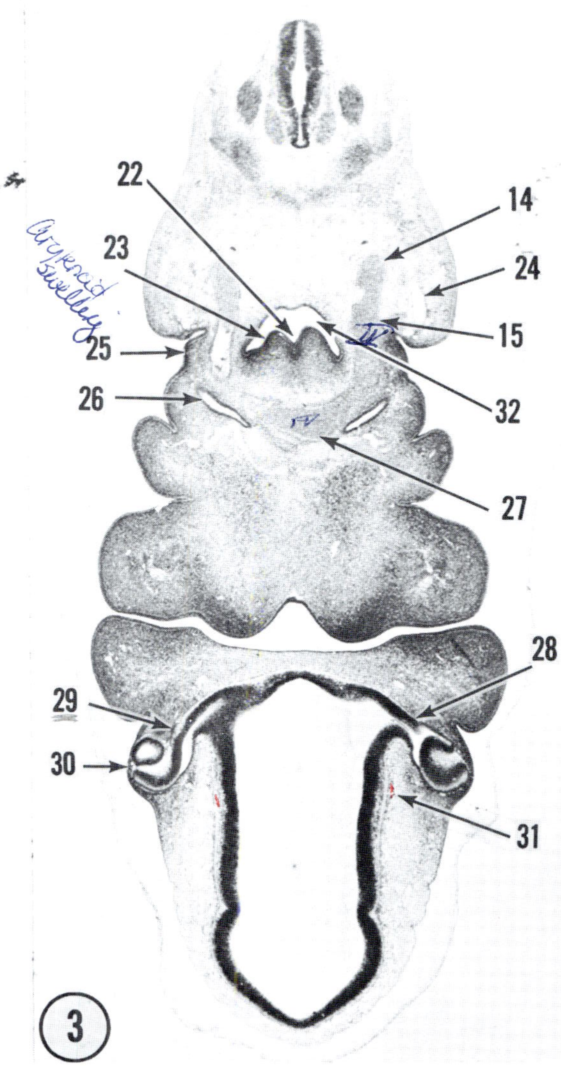

22
14
23
24
15
25
32
26
27
28
29
30
31

3

PLATE 29

10-MM PIG EMBRYO SERIAL TRANSVERSE SECTIONS

1. Dorsal aorta
2. Fifth pharyngeal pouch
3. Fourth pharyngeal pouch
4. Larynx
5. Aortic trunk
6. Stomodeum
7. Diencephalon
8. Third ventricle (major part)
9. Vertebral artery
10. Glottis
11. Pharynx
12. Sixth aortic arch
13. Esophagus
14. Precardinal vein
15. Nasal pit
16. Cerebral hemisphere of telencephalon
17. Third ventricle (minor part)

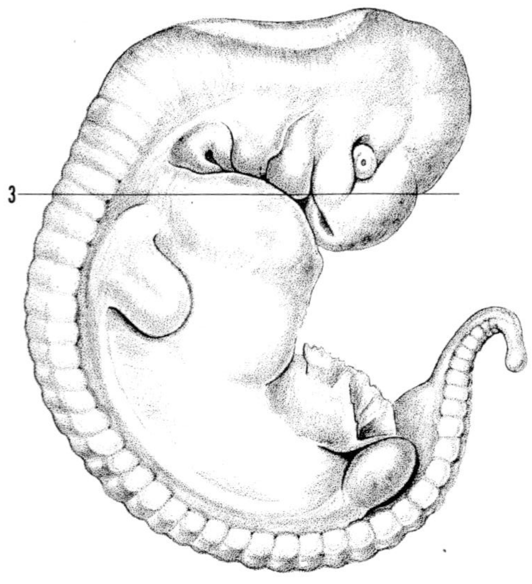

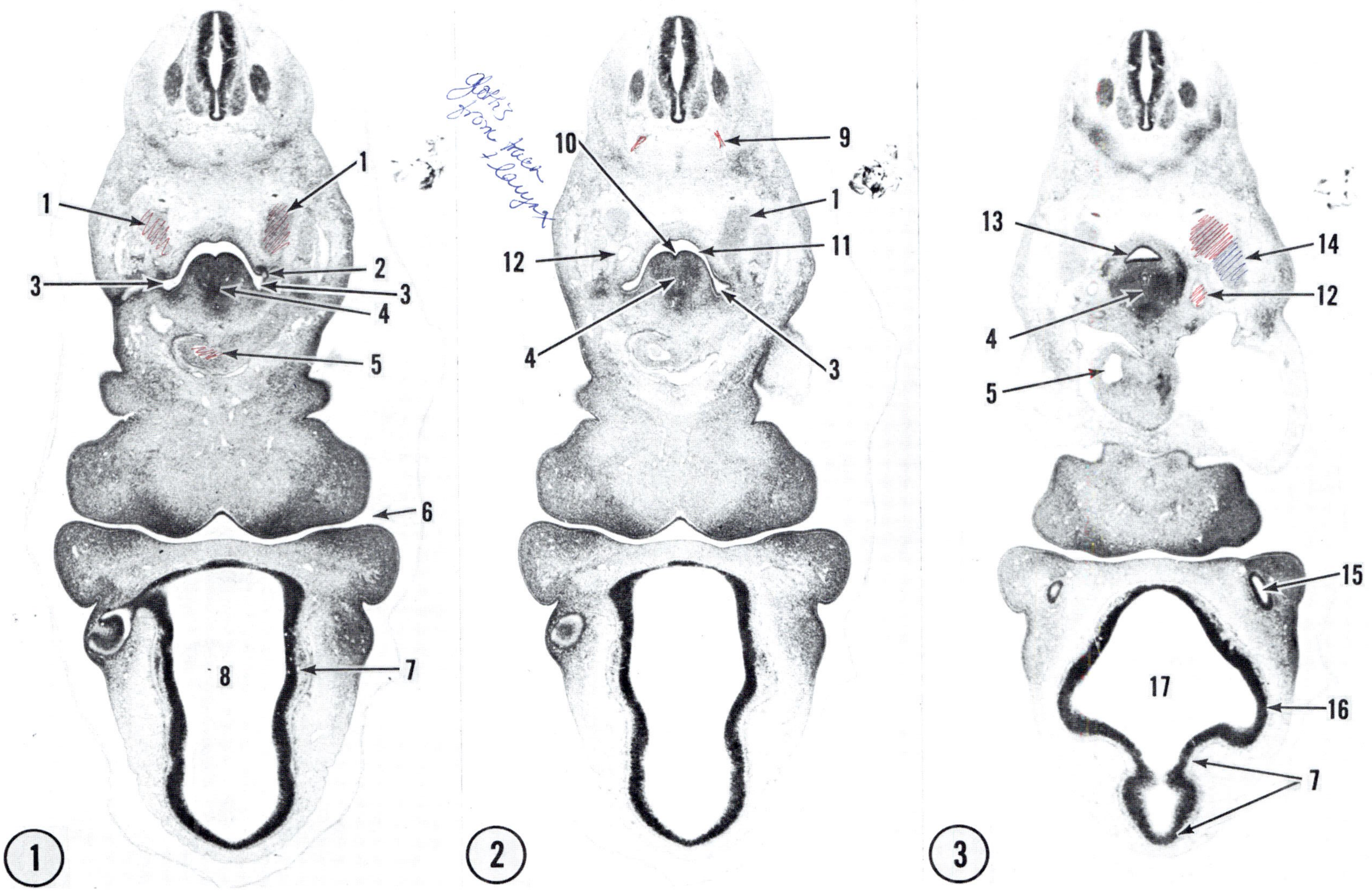

1

2

3

PLATE 30

10-MM PIG EMBRYO SERIAL TRANSVERSE SECTIONS

1. Spinal cord
2. Spinal ganglion
3. Dorsal aorta
4. Precardinal vein
5. Sixth aortic arch
6. Amniotic cavity
7. Aortic trunk
8. Pulmonary trunk
9. Bulbus cordis
10. Bulbar septum
11. Cerebral hemisphere of telencephalon
12. Third ventricle (minor part)
13. Diencephalon
14. Sympathetic chain ganglion
15. Esophagus
16. Trachea
17. Pulmonary artery
18. Left atrium
19. Bulbar ridges
20. Nasal pit
21. Amnion
22. Subclavian vein
23. Right atrium
24. Right ventricle
25. Left ventricle
26. Pericardial cavity
27. Medial nasal process
28. Lateral nasal process
29. Lateral ventricles

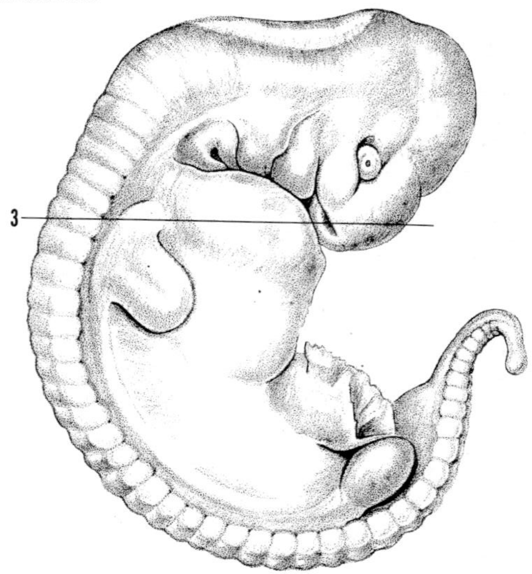

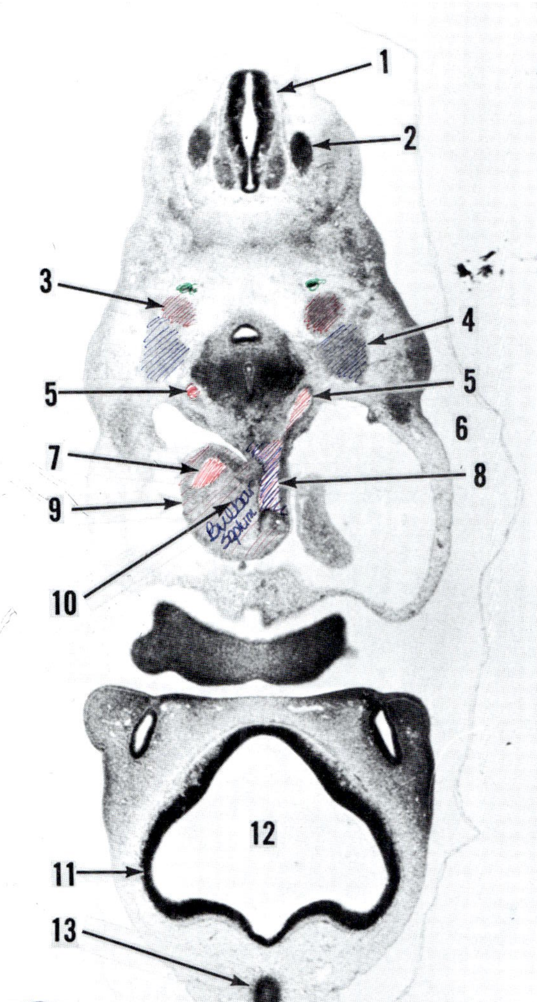
1
2
3
4
5
5
6
7
8
9
10
12
11
13

1

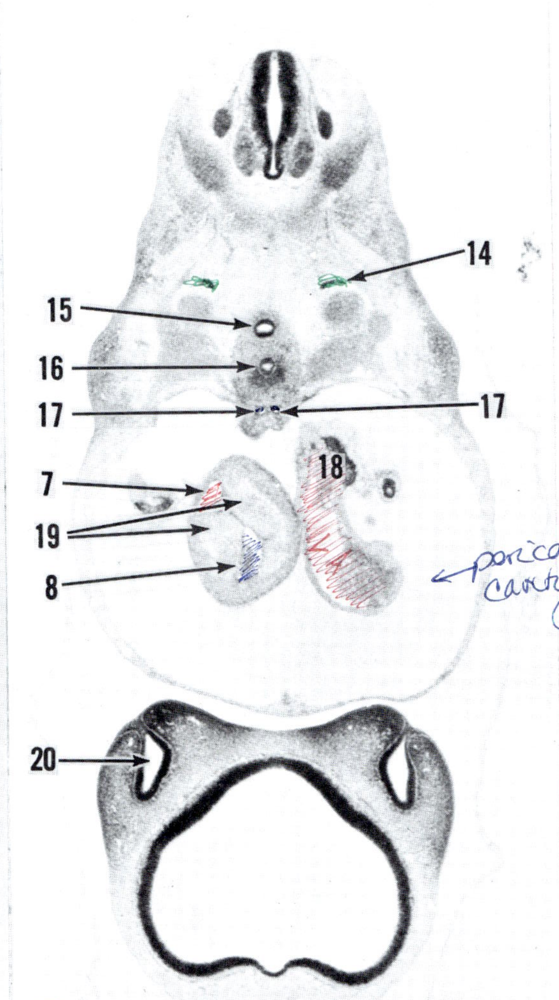
14
15
16
17
17
18
7
19
8
20

2

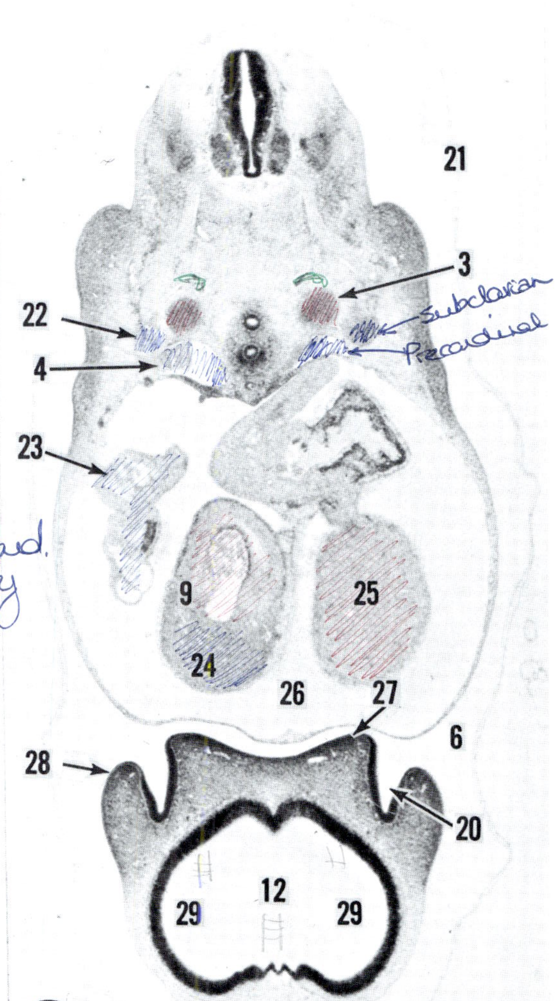
21
3
22
4
23
9
25
24
26
27
6
28
20
12
29
29

3

PLATE 31

10-MM PIG EMBRYO SERIAL TRANSVERSE SECTIONS

1. Subclavian vein
2. Dorsal aorta
3. Common cardinal vein
4. Trachea
5. Foramen secundum
6. Septum primum
7. Left atrium
8. Right atrium
9. Bulbus cordis
10. Right ventricle
11. Nasal pit
12. Cerebral hemisphere of telencephalon
13. Esophagus
14. Vagus nerve
15. Eparterial bronchus
16. Sinus venosus
17. Sino-atrial valve
18. Foramen ovale
19. Endocardial cushion
20. Left ventricle
21. Lamina terminalis
22. Lateral ventricles
23. Intersegmental artery (forms part of adult right subclavian artery)
24. Postcardinal vein
25. Pleural cavity
26. Pulmonary vein continuous with left atrium
27. Atrio-ventricular canal
28. Interventricular foramen
29. Interventricular septum
30. Pericardial cavity
31. Parietal pleura
32. Visceral pleura

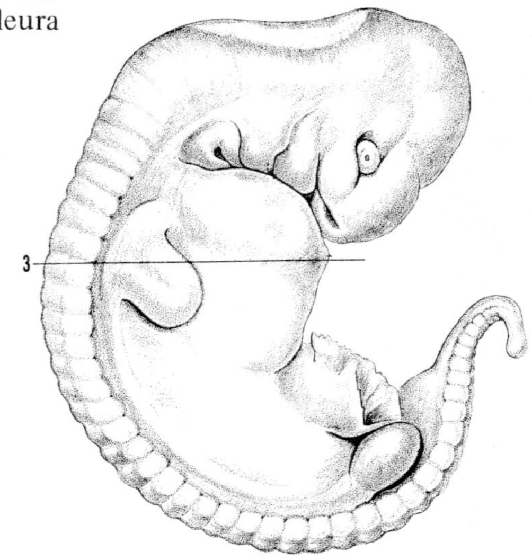

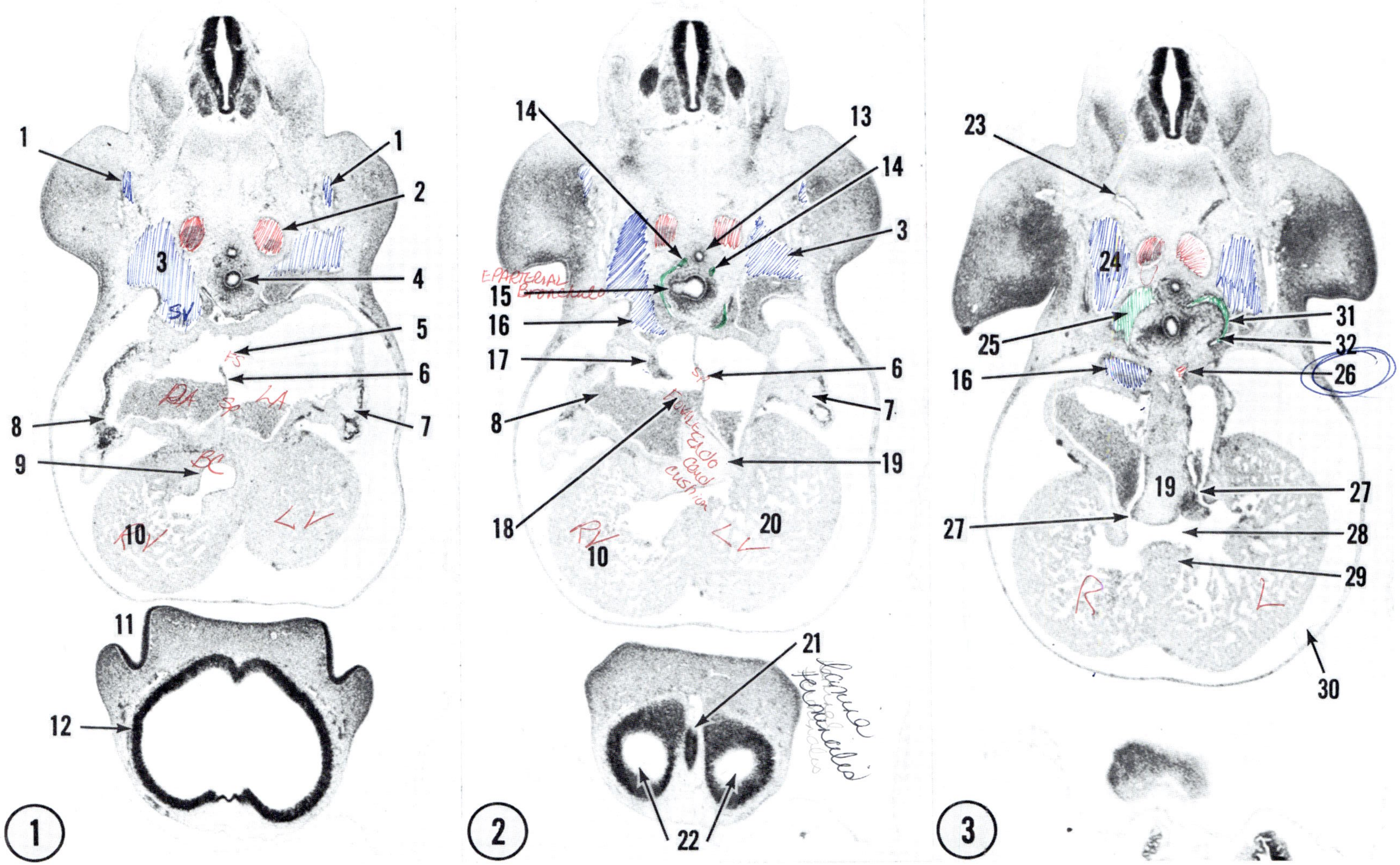
1
2
3
4
5
6
7
8
9
10
11
12
1
13
14
15
16
17
18
19
20
21
22
2
23
24
25
26
27
28
29
30
31
32
3

PLATE 32

10-MM PIG EMBRYO SERIAL TRANSVERSE SECTIONS

1. Spinal cord
2. Spinal ganglion
3. Brachial plexus
4. Postcardinal vein
5. Somatic mesoderm of foreleg bud
6. Esophagus
7. Lung bud
8. Peritoneal cavity
9. Apical ectodermal ridge of foreleg bud
10. Pleural cavity
11. Visceral pleura
12. Septum transversum
13. Inferior vena cava
14. Ventricle
15. Mesoesophagus
16. Descending aorta
17. Mesonephric kidney
18. Left dorsal lobe of liver
19. Left ventral lobe of liver
20. Ductus venosus
21. Yolk sac
22. Dorsal mesogaster
23. Stomach
24. Mammary ridge
25. Omental bursa
26. Gastro-hepatic ligament
27. Peritoneal cavity
28. Hepatic cords
29. Hepatic sinusoids

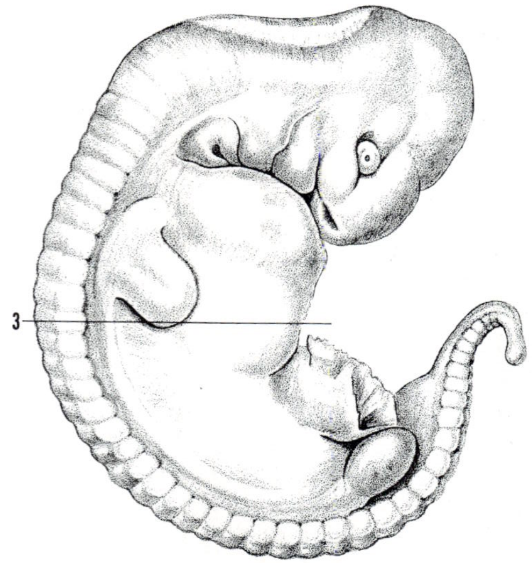

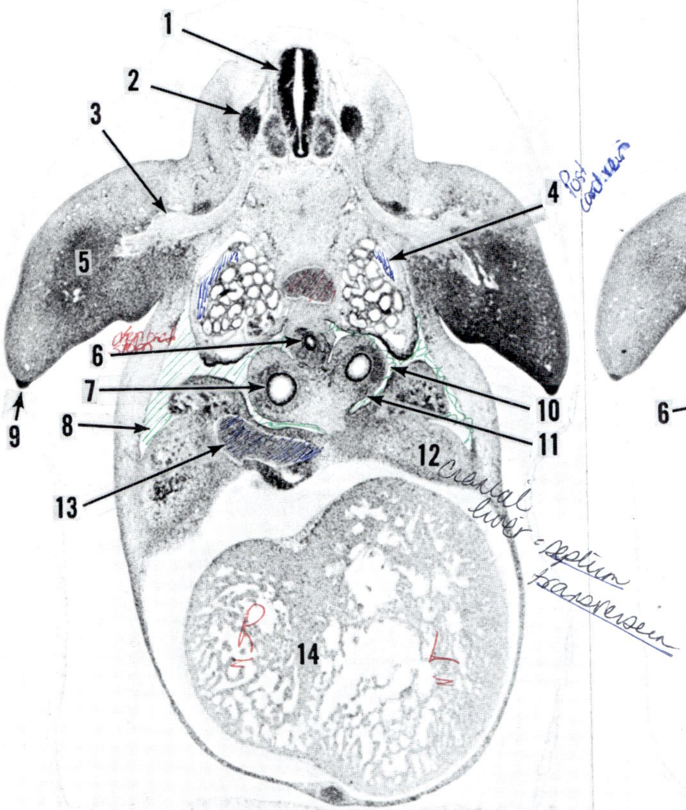
1
2
3
4
5
6
7
8
9
10
11
12
13
14

1

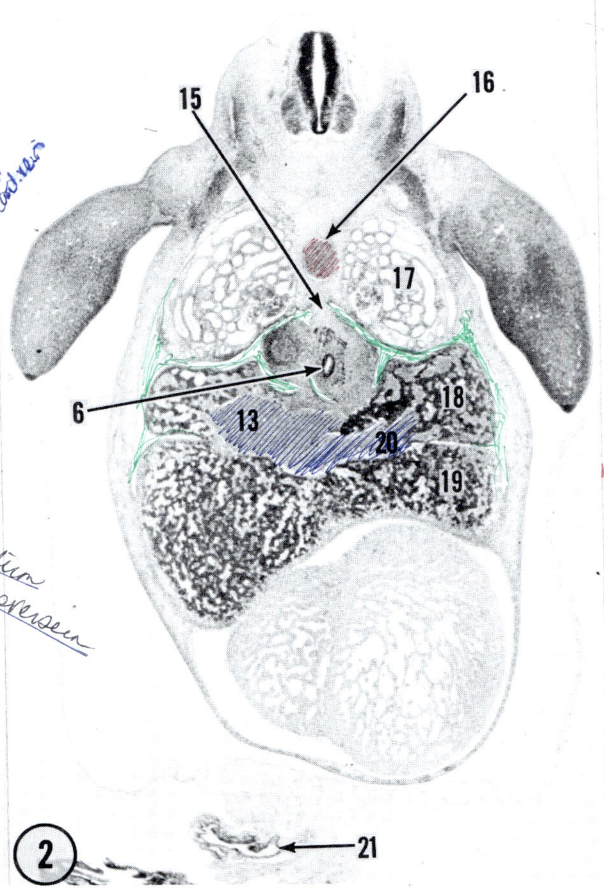
15
16
17
6
13
18
20
19
21

2

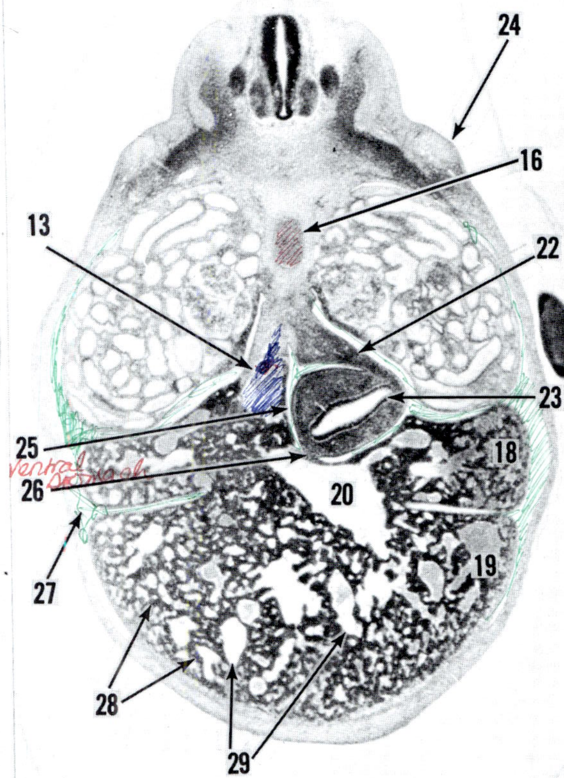
24
16
13
22
23
25
26
27
18
20
19
28
29

3

PLATE 33

10-MM PIG EMBRYO SERIAL TRANSVERSE SECTIONS

1. Descending aorta
2. Mesonephric kidney
3. Inferior vena cava
4. Omental bursa
5. Hepatic portal vein
6. Duodenum
7. Duodeno-hepatic ligament
8. Right dorsal lobe of liver
9. Hepatic duct
10. Right ventral lobe of liver
11. Falciform ligament
12. Temporary umbilical hernia
13. Gonad rudiment
14. Epiploic foramen
15. Parietal peritoneum
16. Visceral peritoneum
17. Common bile duct
18. Cystic duct
19. Right umbilical vein
20. Left umbilical vein
21. Common vitelline vein
22. Cranial limb of the intestinal loop
23. Caudal limb of the intestinal loop
24. Mesonephric tubules
25. Glomeruli
26. Dorsal pancreatic rudiment
27. Germinal epithelium
28. Ventral pancreatic rudiment
29. Dorsal pancreatic duct
30. Gall bladder
31. Hepatic cords
32. Hepatic sinusoid
33. Glomerular capsule

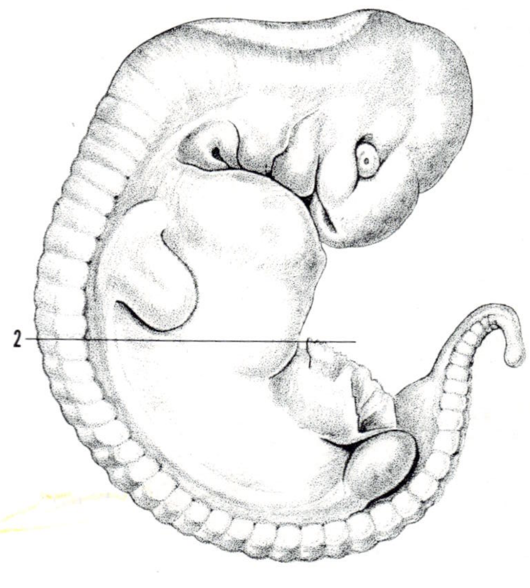

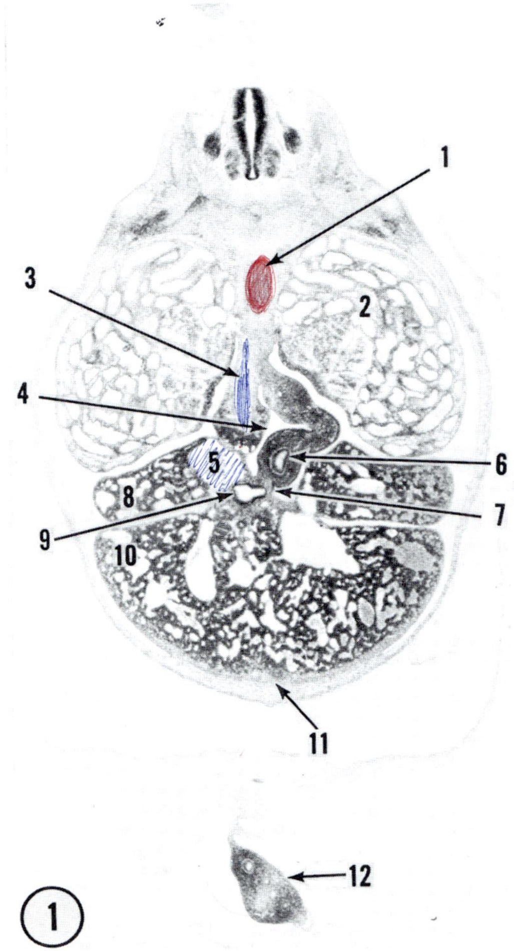
1
2
3
4
5
6
7
8
9
10
11
12

1

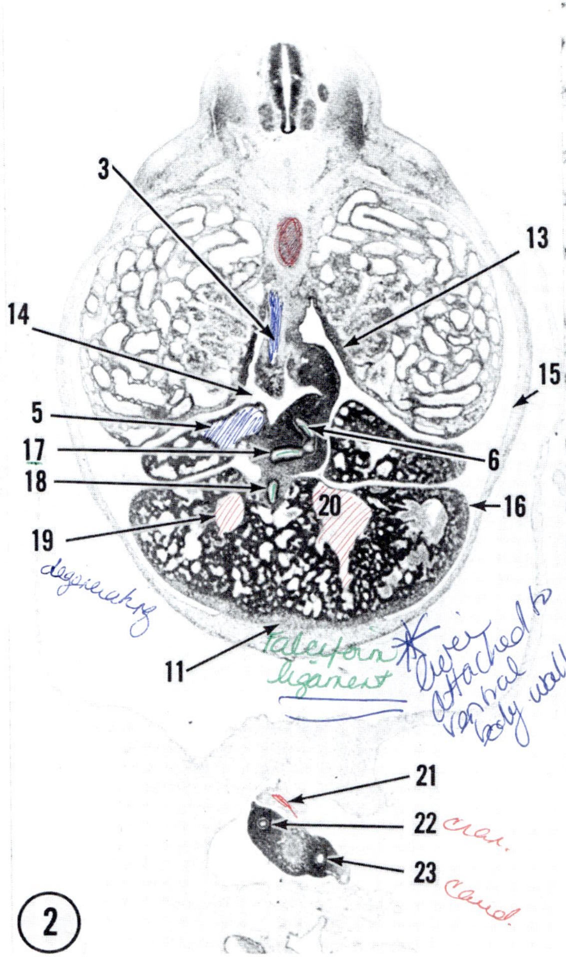
3
13
14
15
5
17
6
18
16
19
20
11
21
22
23

2

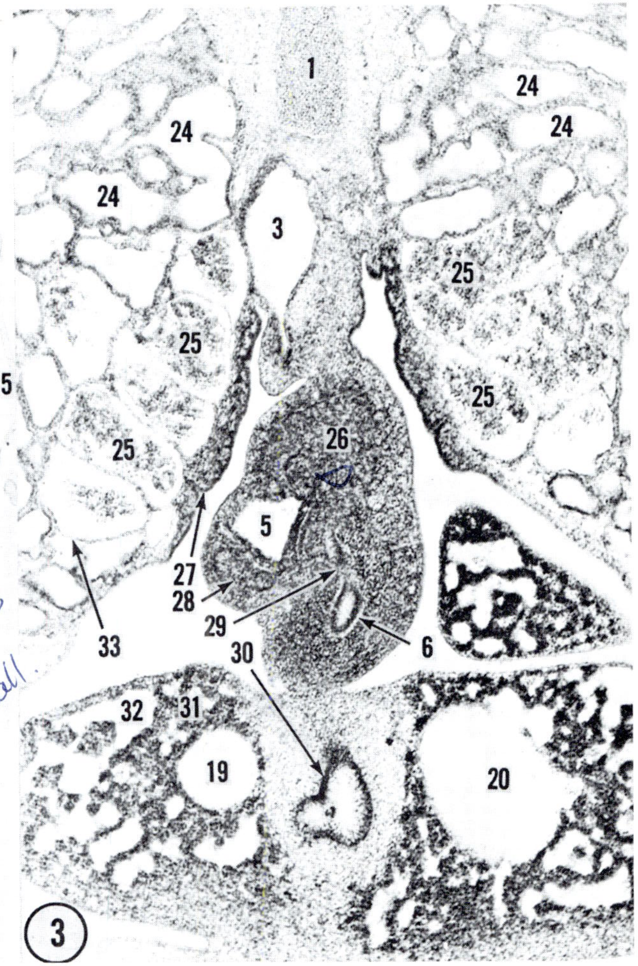
1
24
24
24
24
3
25
25
25
25
26
5
27
28
29
6
33
30
32
31
19
20

3

PLATE 34

10-MM PIG EMBRYO SERIAL TRANSVERSE SECTIONS

1. Descending aorta
2. Mesonephric kidney
3. Subcardinal vein
4. Superior mesenteric vein
5. Duodenum
6. Mesonephric duct
7. Gall bladder
8. Wharton's jelly of umbilical cord
9. Common vitelline vein
10. Umbilical vein
11. Cranial limb of the intestinal loop
12. Superior mesenteric artery
13. Umbilical artery
14. Tail
15. Marginal zone
16. Intermediate zone
17. Ventricular zone
18. Roof plate
19. Alar plate
20. Dorsal root axons
21. Spinal ganglion
22. Basal plate
23. Motor horn
24. Dorsal root dendrites
25. Spinal nerve – dorsal ramus
26. Ventral root
27. Floor plate
28. Notochord
29. Sclerotome
30. Spinal nerve – ventral ramus
31. Spinal nerve – visceral ramus
32. Sympathetic chain ganglion
33. Mesonephric tubules
34. Gonad rudiment
35. Caudal limb of the intestinal loop
36. Allantois

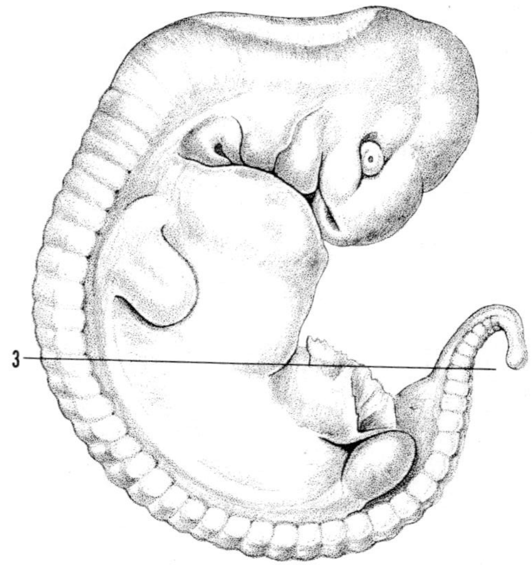

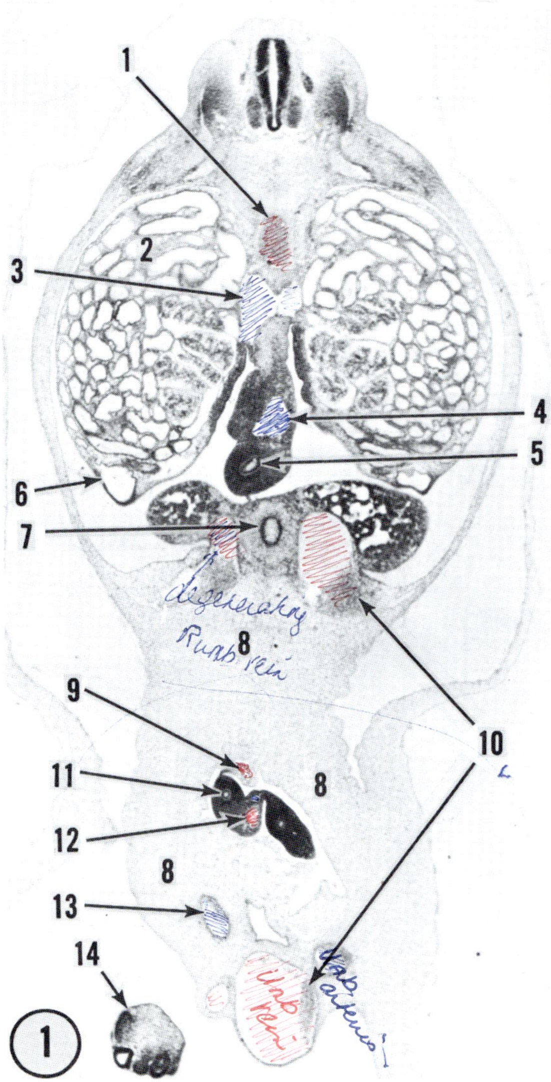
1
2
3
4
5
6
7
8
9
10
11
12
13
14

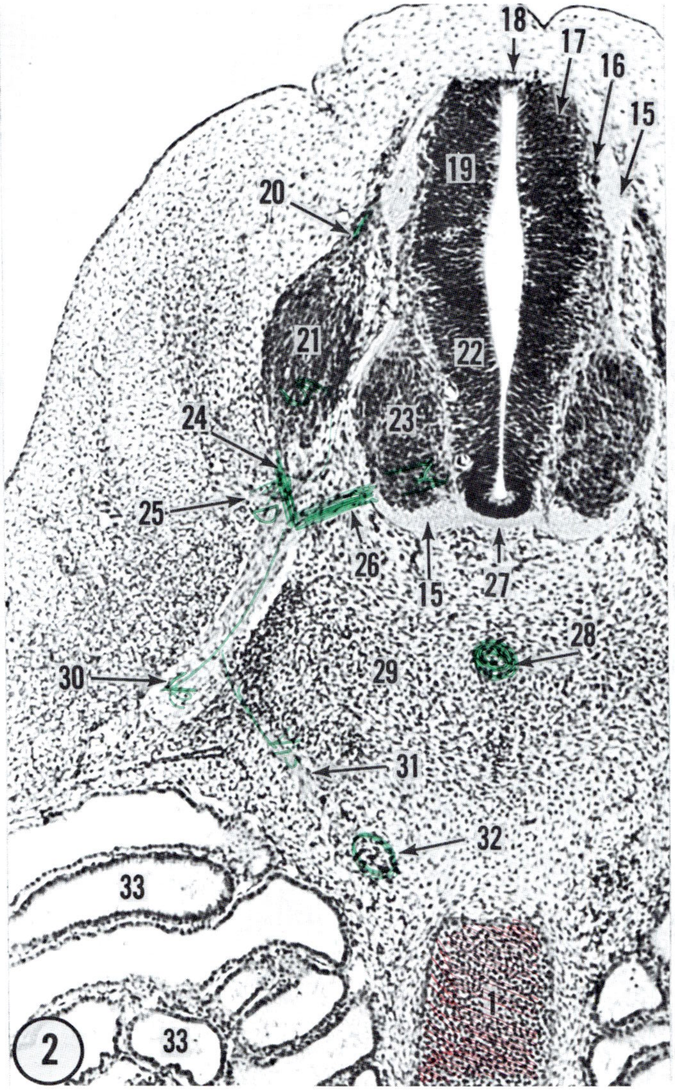
15
16
17
18
19
20
21
22
23
24
25
26
27
28
29
30
31
32
33

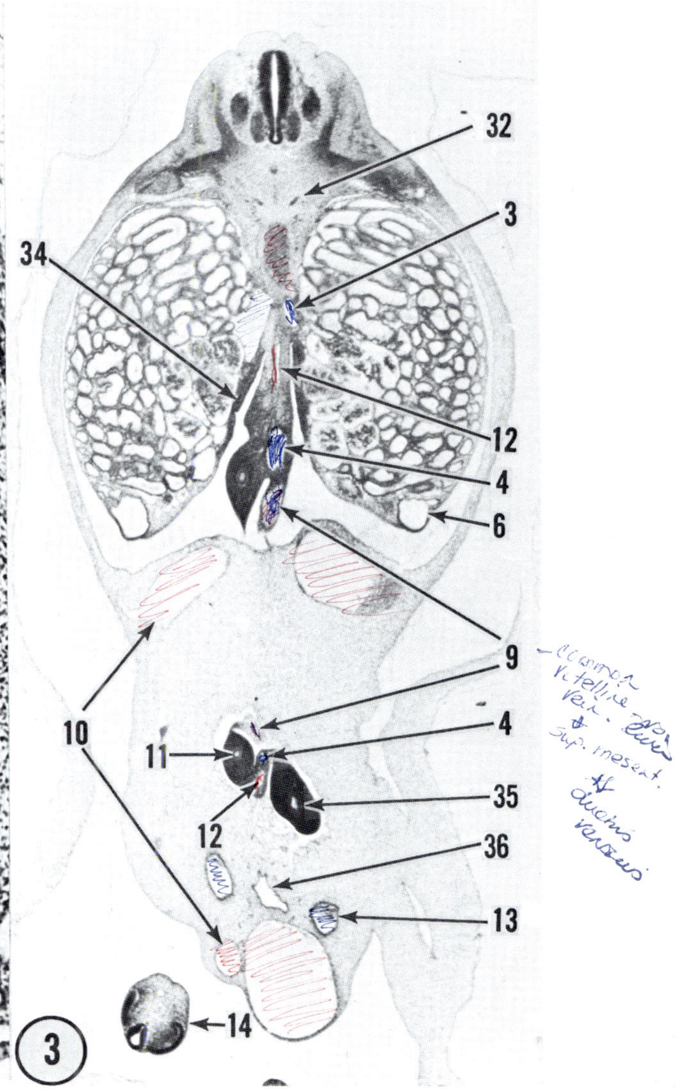
3
4
6
9
10
11
12
13
14
32
34
35
36

PLATE 35

10-MM PIG EMBRYO SERIAL TRANSVERSE SECTIONS

1. Mesonephric tubules
2. Subcardinal anastomosis
3. Glomerular capsule
4. Mesonephric duct
5. Continuity between the duodenum (above arrow) and cranial limb of the intestinal loop (below arrow)
6. Superior mesenteric vein
7. Caudal limb of the intestinal loop
8. Umbilical vein
9. Umbilical artery
10. Tail
11. Extra-embryonic coelom
12. Descending aorta
13. Glomeruli
14. Gonad rudiment
15. Superior mesenteric artery
16. Mammary ridge
17. Genital eminence
18. Cloacal membrane
19. Tail gut
20. Mesocolon
21. Caudal limb of the intestinal loop (forms part of adult colon)
22. Allantois
23. Peritoneal cavity
24. Cloaca (future urogenital sinus)
25. Cloacal septum
26. Cloaca (future rectum)
27. Spinal cord
28. Mesonephric kidney

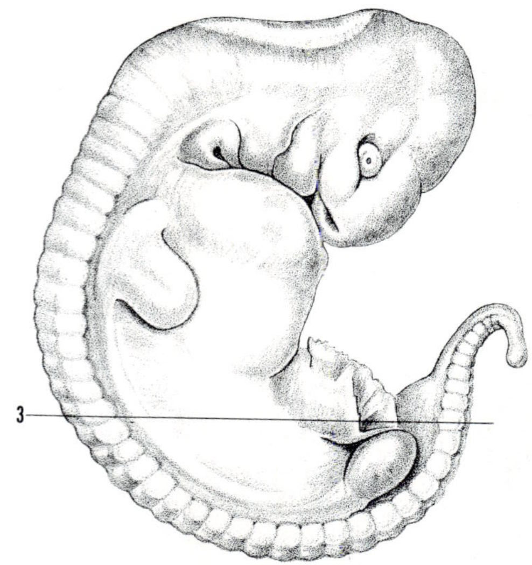

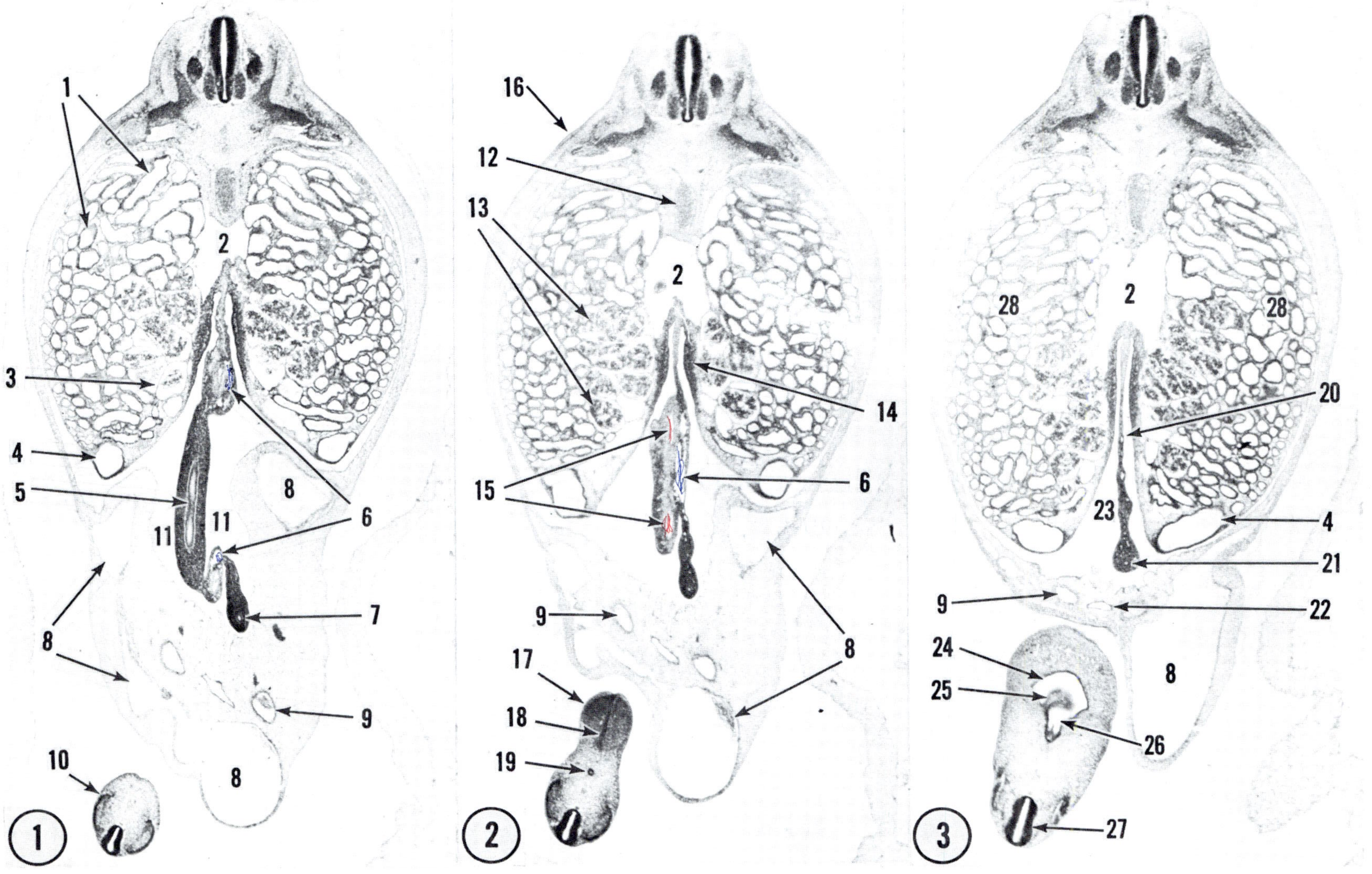
1
2
3
4
5
6
7
8
9
10
11
12
13
14
15
16
17
18
19
20
21
22
23
24
25
26
27
28

PLATE 36

10-MM PIG EMBRYO SERIAL TRANSVERSE SECTIONS

1. Subcardinal anastomosis
2. Mesocolon
3. Umbilical artery
4. Allantois (continuous below with urogenital sinus)
5. Umbilical vein
6. Mesonephric duct
7. Urogenital sinus
8. Colon
9. Descending aorta
10. Mesonephric kidney
11. Caudal limb of the intestinal loop
12. Peritoneal cavity
13. Spinal cord
14. Motor horn
15. Spinal ganglion
16. Mammary ridge
17. Apical ectodermal ridge of hindleg bud
18. Somatic mesoderm of hindleg bud
19. Ureters
20. Postcardinal vein
21. Caudal artery
22. Notochord

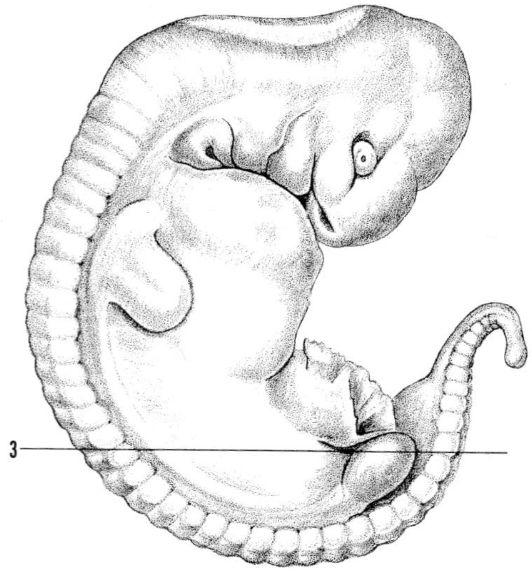

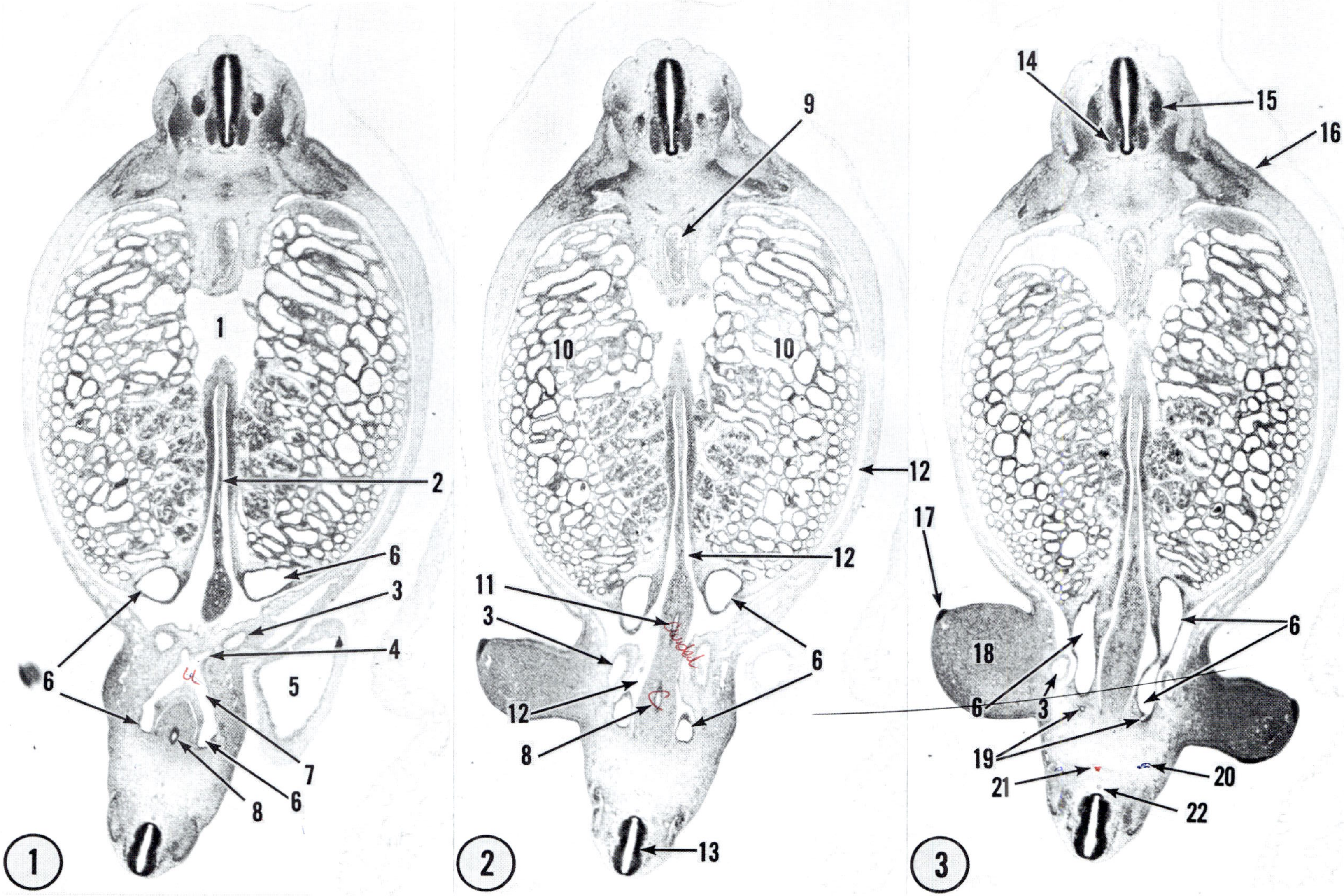

1 2 3

PLATE 37

10-MM PIG EMBRYO SERIAL TRANSVERSE SECTIONS

1. Umbilical artery
2. External iliac artery
3. Renal pelvis
4. Caudal artery
5. Nephrotome
6. Descending aorta
7. Mesonephric kidney
8. Peritoneal cavity
9. Postcardinal vein
10. Notochord
11. Spinal ganglion
12. Spinal cord
13. Skin ectoderm
14. Myotome
15. Dermatome
16. Ventral root
17. Sclerotome – dense caudal portion
18. Sclerotome – less dense cranial portion
19. Cavity of the spinal cord

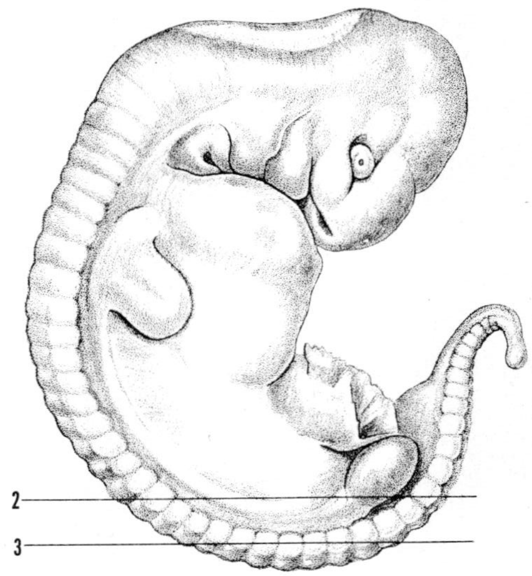

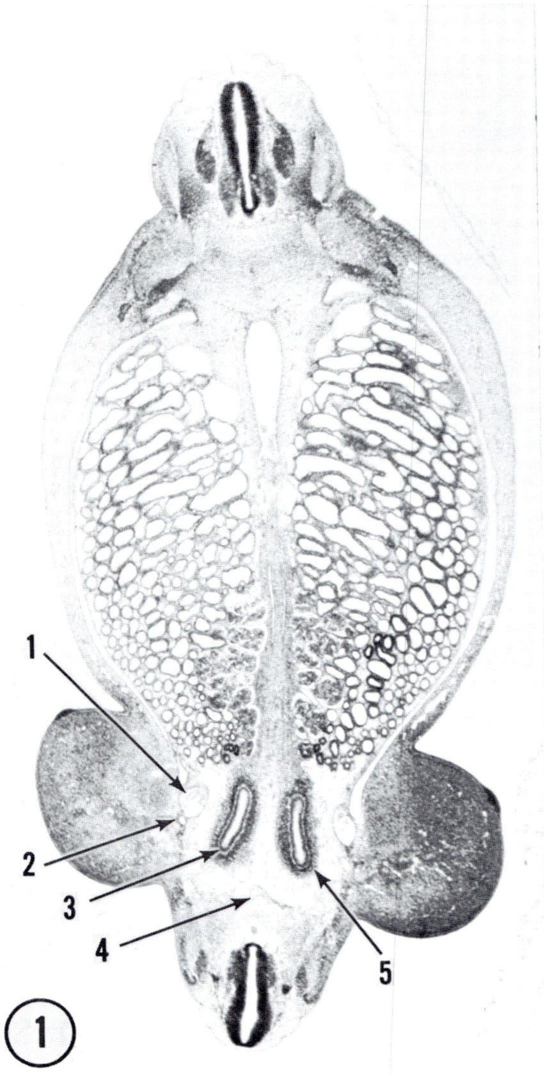

1

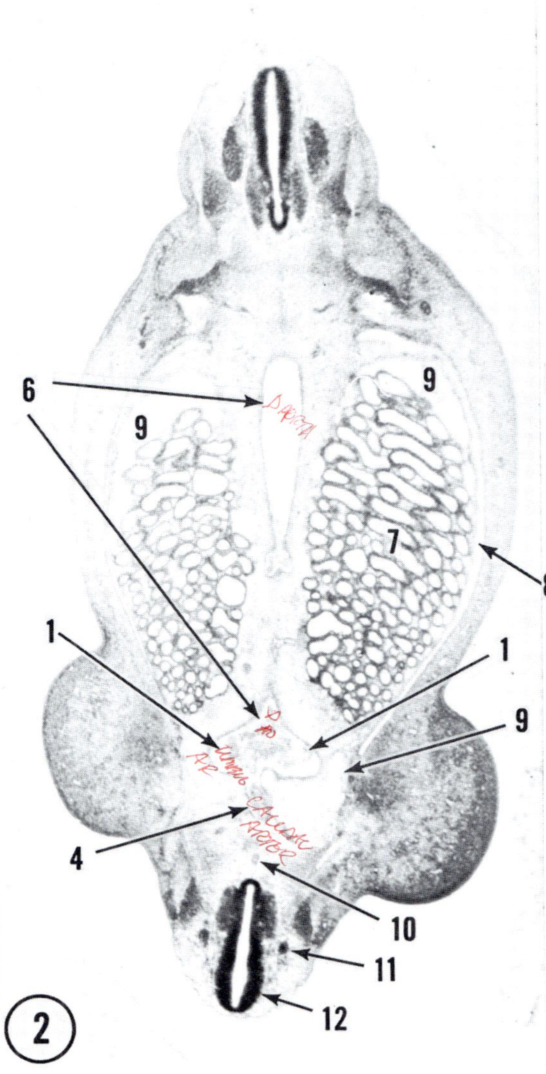

2

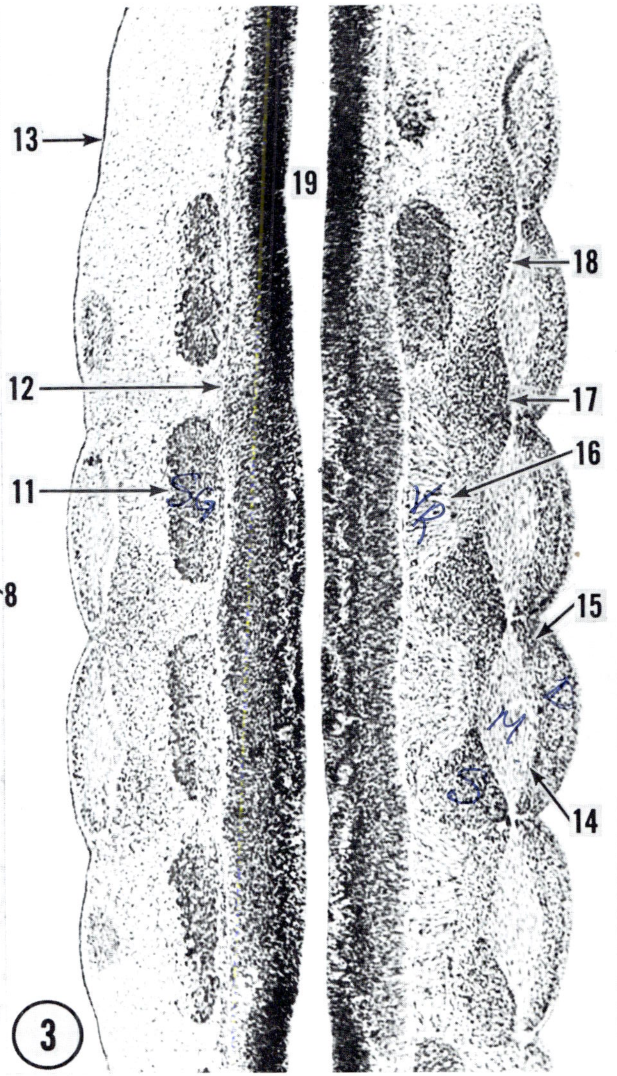

3

PLATE 38

FROG: BLASTULA, GASTRULAE, AND NEURULA

FIGURE 1 Blastula (sagittal section)

FIGURE 2 Dorsal lip gastrula (sagittal section)

FIGURE 3 Yolk plug gastrula (sagittal section)

FIGURES 4, 5 Neural plate neurula (transverse section)

(Figure 5 is an enlargement of dorsal portion of Figure 4)

AP Animal pole
D Dorsal side
V Ventral side
VP Vegetal pole

1. Pigmented cortex
2. Vitelline membrane
3. Blastocoel
4. Area of gray crescent
5. Shrinkage spaces
6. Nuclei
7. Blastomeres
8. Outer ectodermal layer
9. Inner ectodermal layer
10. Archenteron roof (mesoderm)
11. Direction of epiboly
12. Dorsal blastoporal lip
13. Blastopore
14. Archenteron
15. Yolk-filled endodermal cells
16. Yolk plug
17. Ventral blastoporal lip
18. Neural plate
19. Skin ectoderm
20. Lateral plate mesoderm
21. Directions of cellular migration to form endodermal roof of archenteron
22. Segmental plate mesoderm
23. Notochord
24. Archenteron roof (endoderm)

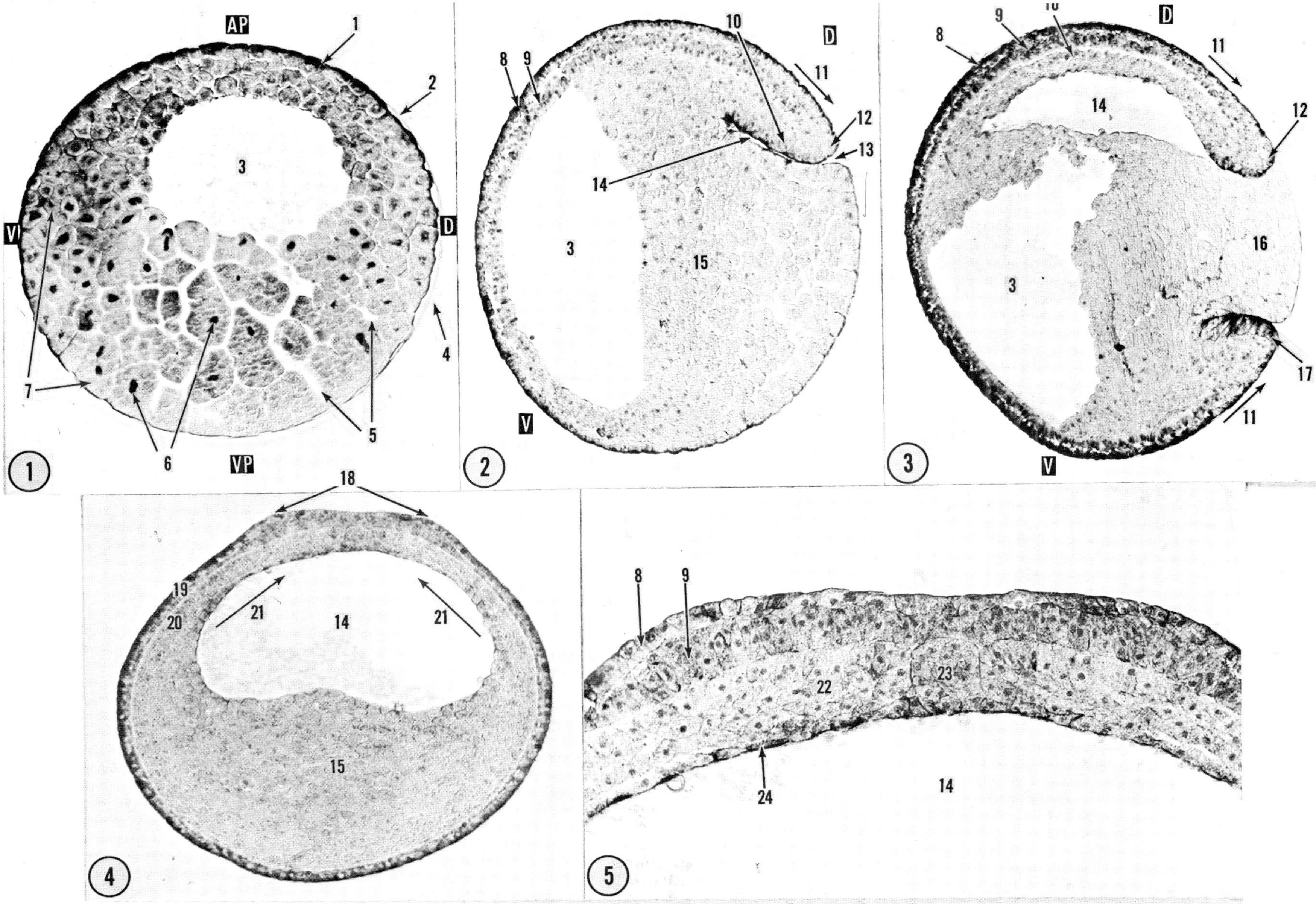
AP
VP
V
D
V
D
V
D
1
2
3
4
5
6
7
8
9
10
11
12
13
14
15
16
17
18
19
20
21
22
23
24
1
2
3
4
5

PLATE 39

FROG: NEURULA AND 4-MM EMBRYO SERIAL TRANSVERSE SECTIONS

FIGURES 1, 2 Neural fold neurula (transverse section)

FIGURES 3, 4 4-mm embryo serial transverse sections

(Figure 2 is an enlargement of dorsal portion of Figure 1)

1. Neural groove
2. Skin ectoderm
3. Lateral plate mesoderm
4. Archenteron
5. Yolk-filled endodermal cells
6. Neural fold
7. Outer ectodermal layer
8. Inner ectodermal layer
9. Segmental plate mesoderm
10. Notochord
11. Archenteron roof (endoderm)
12. Pineal gland
13. Prosencephalon
14. Nasal placode
15. Mesencephalon
16. Pigmented retina
17. Sensory retina
18. Optic stalk
19. Rudiment of the anterior pituitary gland
20. Head mesenchyme
21. Stomodeum

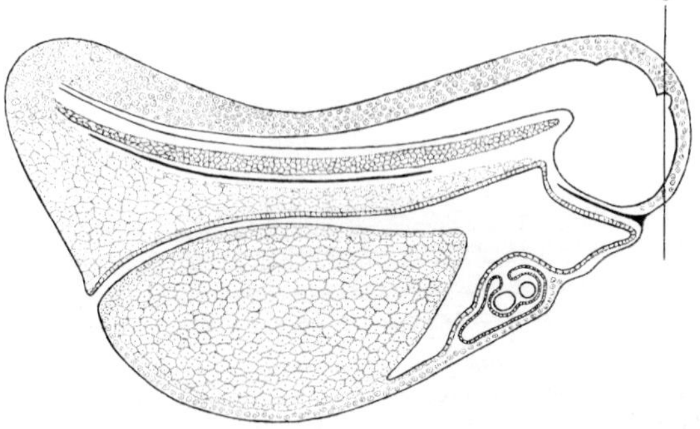

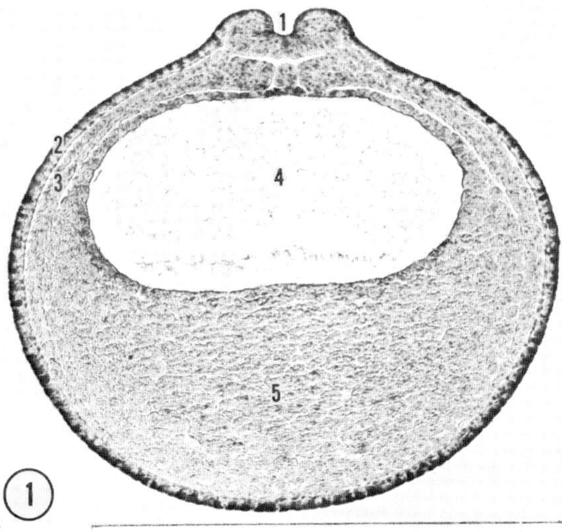

1

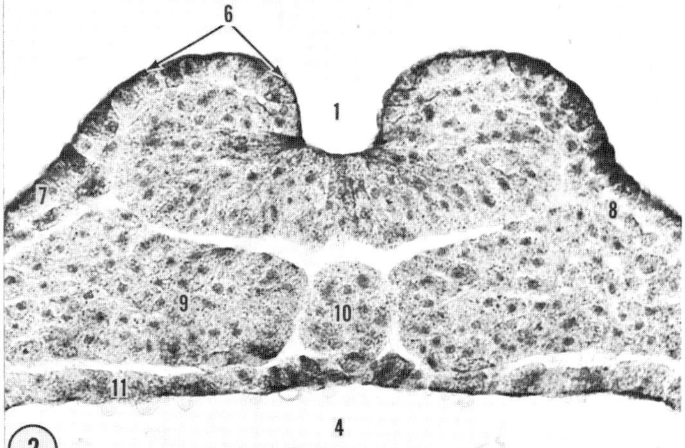

2

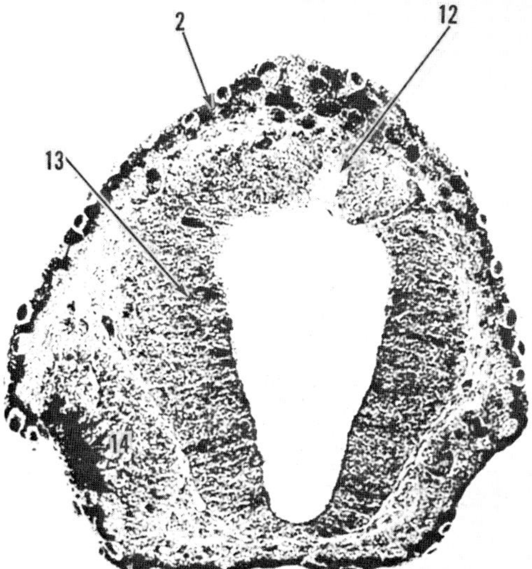

3

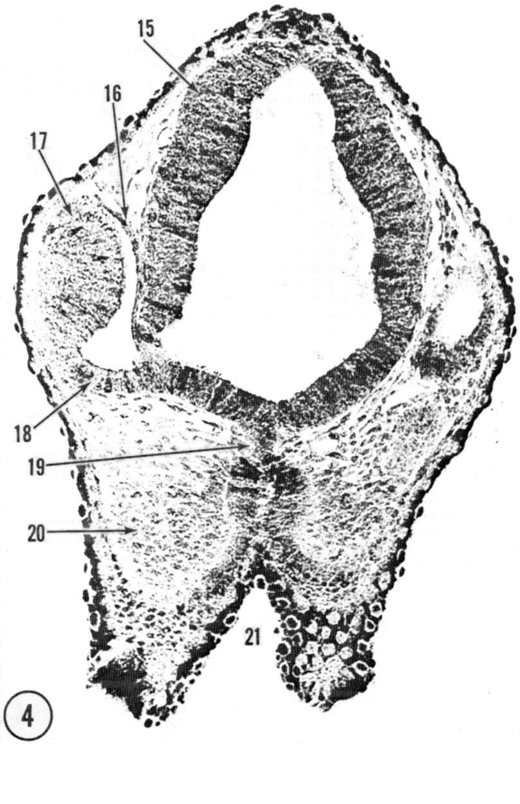

4

PLATE 40

4-MM FROG EMBRYO SERIAL TRANSVERSE SECTIONS

1. Mesencephalon
2. Future corneal epithelium
3. Lens placode
4. Rudiment of the anterior pituitary gland
5. Foregut
6. Oral membrane
7. Adhesive gland
8. Stomodeum
9. Rhombencephalon
10. Skin ectoderm
11. Semilunar ganglion
12. Infundibulum
13. Auditory vesicle
14. Notochord
15. Dorsal aorta
16. Pharynx
17. First aortic arch
18. Bulbus cordis
19. Duodenum
20. Liver rudiment
21. Dorsal mesocardium
22. Pericardial cavity
23. Endocardium of ventricle
24. Epi-myocardium of ventricle
25. Parietal pericardium

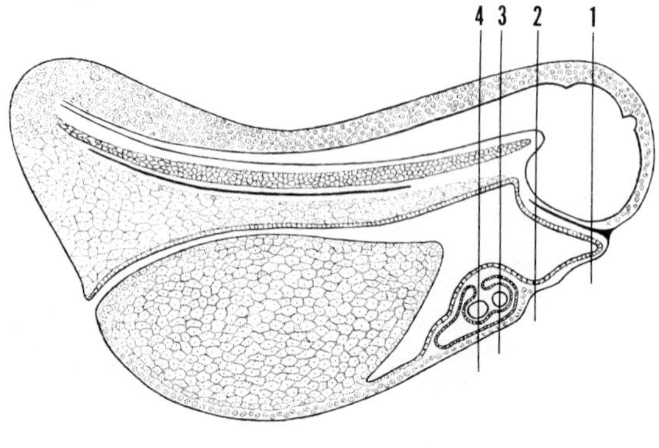

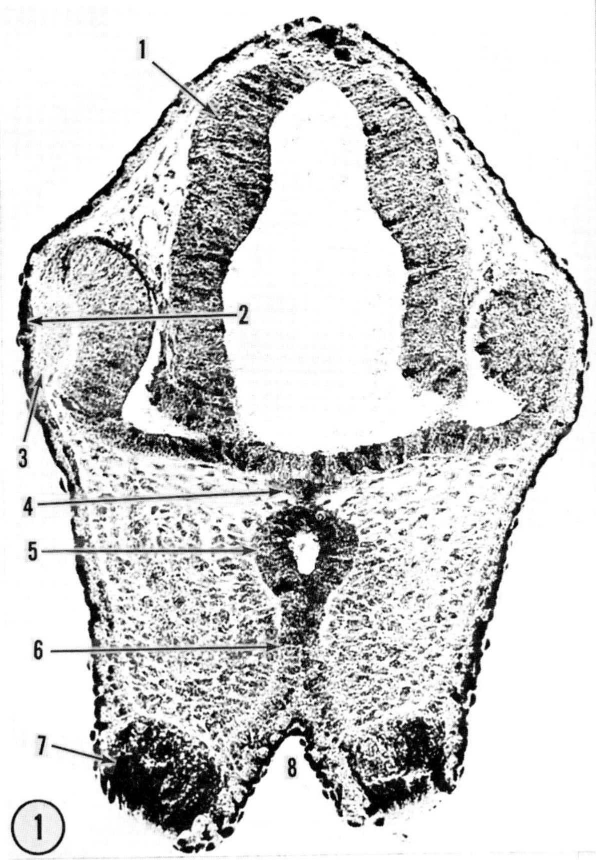

1

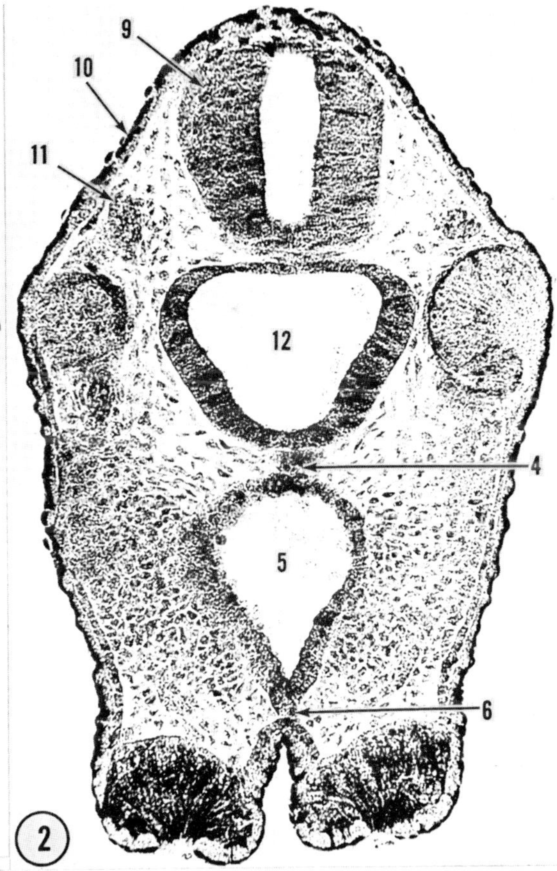

2

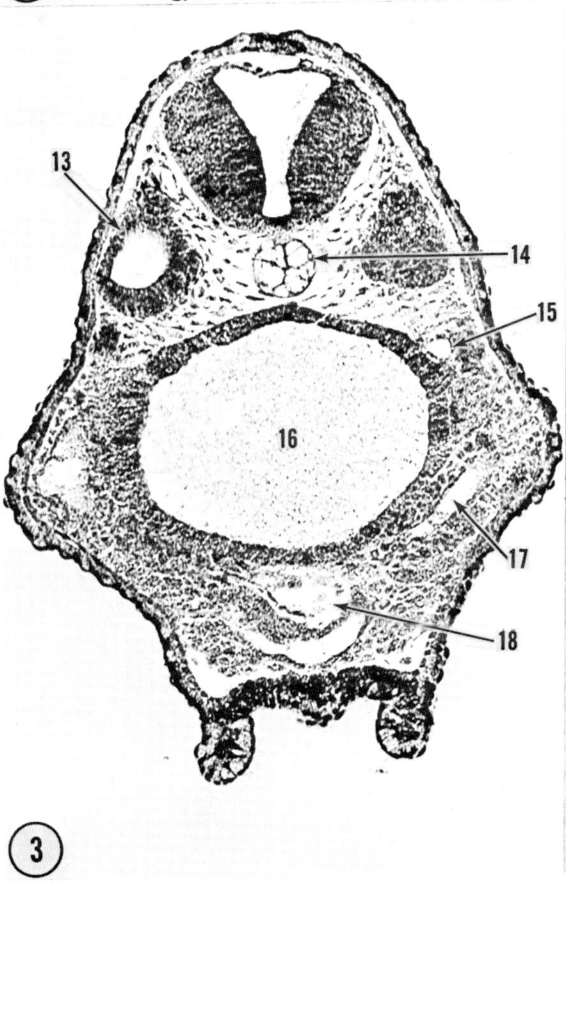

3

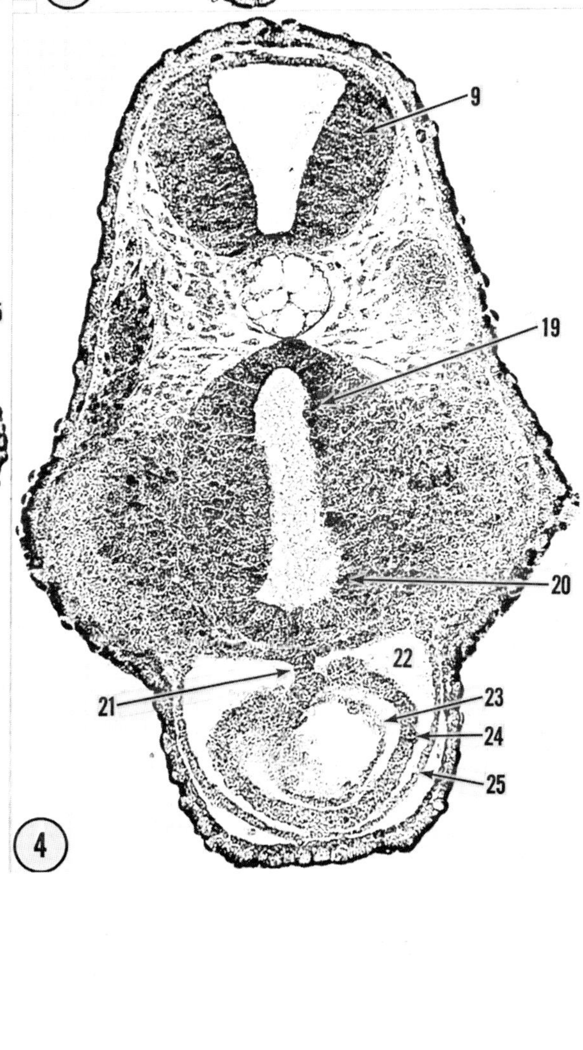

4

PLATE 41

4-MM FROG EMBRYO SERIAL TRANSVERSE SECTIONS

1. Roof plate
2. Neural crest cells
3. Spinal cord
4. Floor plate
5. Dorsal aorta
6. Pronephric tubule
7. Pronephric duct
8. Pronephric ridge
9. Midgut
10. Yolk-filled endodermal cells
11. Skin ectoderm
12. Liver rudiment
13. Dorsal fin
14. Somite
15. Notochord
16. Subnotochordal rod
17. Hindgut
18. Proctodeum
19. Ventral fin

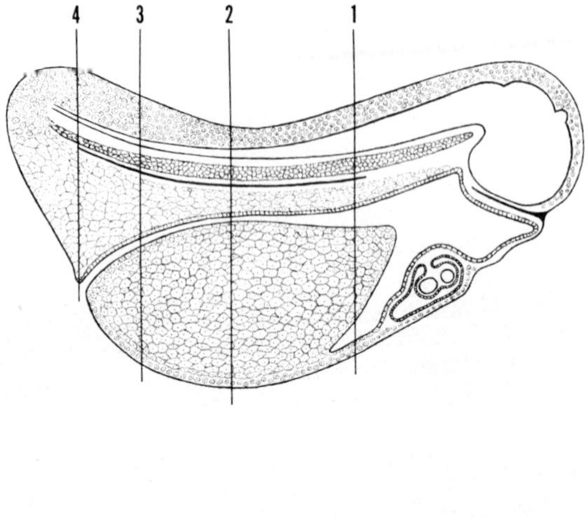

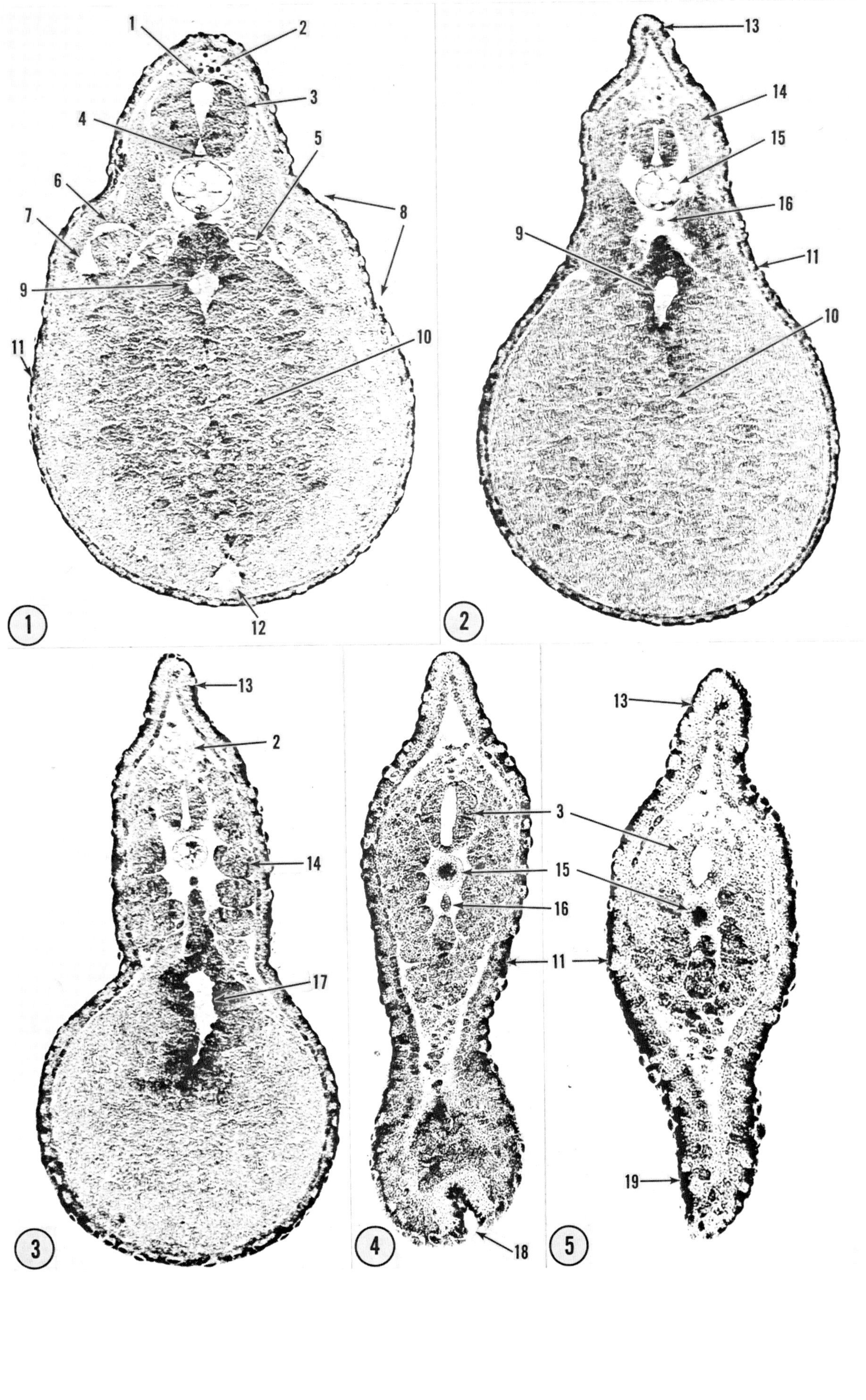
1
2
3
4
5
6
7
8
9
10
11
12
13
14
15
16
17
18
19

Abridged Index

(Contains every term listed in text in ALL CAPITAL LETTERS)